SHATTERED CONSENSUS

SHATTERED CONSENSUS

The True State of Global Warming

Edited by
PATRICK J. MICHAELS

ROWMAN & LITTLEFIELD PUBLISHERS, INC.
Lanham • Boulder • New York • Toronto • Oxford

ROWMAN & LITTLEFIELD PUBLISHERS, INC.

Published in the United States of America
by Rowman & Littlefield Publishers, Inc.
A wholly owned subsidiary of The Rowman & Littlefield Publishing Group, Inc.
4501 Forbes Boulevard, Suite 200, Lanham, Maryland 20706
www.rowmanlittlefield.com

PO Box 317
Oxford
OX2 9RU, UK

363.7387

British Library Cataloging in Publication Information Available

Library of Congress Cataloging-in-Publication Data

Shattered consensus : the true state of global warming / edited by Patrick J.
Michaels.
    p. cm.
Includes bibliographical references and index.
ISBN 0-7425-4922-4 (alk. paper) — ISBN 0-7425-4923-2 (pbk. : alk. paper)
    1. Climatic changes—Research. 2. Global warming—Research. 3. Greenhouse
effect, Atmospheric. 4. Greenhouse gases—Environmental aspects. I. Michaels,
Patrick J.
    QC981.8.C5S5 2005
    363.738'74'072—dc22   2005006358

Printed in the United States of America

♾ ™ The paper used in this publication meets the minimum requirements of
American National Standard for Information Sciences—Permanence of Paper for
Printed Library Materials, ANSI/NISO Z39.48-1992.

CONTENTS

FOREWORD

It is the tension between creativity and skepticism that has
produced the stunning and unexpected findings of science.

—Carl Sagan

No science is immune to the infection of politics and the cor-
ruption of power.

—Jacob Bronowski

The strongest arguments prove nothing so long as the conclusions are not verified by experience. Experimental science is
the queen of sciences and the goal of all speculation.

—Roger Bacon

With few exceptions, scientific controversies rarely intrude into the
consciousness of the general public, yet the vocal debates, in America and worldwide, over climate change are widely reported. The reasons for
this are not surprising: weather affects everyone, every day, and the social and
economic impacts of a radically changed climate are hard to comprehend.

In fact, people have little understanding of the exact nature and causes
of climate change, in spite of—or perhaps because of—the vast amount of
sensational literature available. There are many reasons for this. Specialists
often have difficulty explaining their research to the public, in part because
the basic principles of scientific methodology are not widely known. Even
fundamental principles of climatology—such as that climate *always* changes,
sometimes very abruptly in as little as a few decades—are unfamiliar to most

people. The mechanisms and pace of climate change remain the subject of considerable uncertainty and scientific inquiry. Given that, it is not surprising that the public has a difficult time making sense of the debate.

As the essays in this volume reveal, the state of certainty about the science of climate change is greatly exaggerated. Major questions have yet to be adequately resolved, including natural variability; climate sensitivity, which is the effect on temperature from a doubling of CO_2; cloud formation; water vapor; climate feedbacks; temperature differences between the surface and lower troposphere; ocean currents; aerosols; and solar and lunar influences. The inability to fully understand these complex systems has significant implications for public policy. Yet this message is rarely carried to the public. And when it is, those with the fortitude to raise such questions are dismissed as skeptics or worse.

The climate change issue burst on the public scene in 1989. During a second consecutive, unusually hot summer, a NASA scientist testified before Congress that global warming was taking place and was serious. With that, the apocalyptic bandwagon got rolling. No one paused to consider that just twelve years earlier the National Science Board told the new Department of Energy that the "present time of high temperatures should be drawing to an end . . . leading into the next glacial age." The fact that scientific opinion could change so radically in little over a decade should have stopped the rush to judgment. And it should have certainly served as a healthy dose of humility. It did neither. Our greatest problem is not ignorance; it is the presumption of knowledge.

A powerful and hardened orthodoxy contending that the "science is settled and that there is a consensus" among scientists that human activities are causing serious ecological problems clearly has formed. These tenets of faith are usually included in skillfully worded statements that imply more than they actually say. Further, the notion of a scientific consensus shapes public perceptions and pushes policy choices toward the acceptance of constraints on energy use.

The influence of this imagery is clearly evident. The periodic summations of the state of climate science by the Intergovernmental Panel on Climate Change (IPCC) carry enormous weight in policy circles around the globe, shape how the media treats the climate issue, and drive the Kyoto Protocol internationally and calls for a cap-and-trade program or renewable portfolio standards in the United States.

But the science on climate change is far from settled, and no one can show with analytical rigor that climate change over the course of this century will be the cause of serious ecological damage.

Since climate always changes, it is the ultimate apocalyptic issue. Freeze or fry, wet or dry, industrial activity is the culprit, and mandates controlling economic activity are the solution. Alarmism is a proven technique for fund raising, gaining recognition, and rushing to judgment.

Alarmism finds strength in our national susceptibility to being misled. This was insightfully documented four decades ago by the eminent historian Daniel Boorstin in his book *The Image: A Guide to Pseudo-Events in America*. Boorstin made the point that there was a growing gap between what an informed citizen needs to know and can know. This gap, combined with extravagant expectations, makes us susceptible to being misled and to self-deception. We have created a world where reality is tested by images, rather than images being tested by reality. Regrettably, it has become easier to exploit problems than solve them.

The essays in this volume expose the lack of certainty in the bold statements of fact by the IPCC and document the numerous misstatements, omissions, and mistaken conclusions of the Third Assessment's Summary for Policymakers (SPM). The systematic overstatement of the interpretative findings of the SPM should be enough to give pause and to encourage greater awareness of the actual details of the scientific record. This volume offers those details so readers can decide for themselves whether the alarmism about climate change is justified.

Policy makers will be better served if they know where there is legitimate consensus and where there are major disagreements. Clarity in communication could begin by articulating what is known about human influence, what is unknown about the climate system but perhaps knowable in a reasonable time with additional research, what may be unknowable in any reasonable time period, what information is the most important for near-term policy decisions, and why specific uncertainties are important for policy making.

Wise, effective climate policy flows from a sound scientific foundation, a clear understanding of what science can and cannot tell us about human influence on the climate system, and the consequences of alternative courses of action to manage risk.

—William O'Keefe, CEO
Jeffrey Kueter, President
The George C. Marshall Institute

1

INTRODUCTION

FALSE IMPRESSIONS: MISLEADING STATEMENTS, GLARING OMISSIONS, AND ERRONEOUS CONCLUSIONS IN THE IPCC's SUMMARY FOR POLICYMAKERS, 2001

Patrick J. Michaels

By its very nature, science is an exercise that is never finished. Consequently, any attempt to summarize the "state" of any branch of science, such as climate science, will always be one or many steps behind the true configuration of that particular area of study. Comprehensive compilations of climate science, such as those of the U.N. Intergovernmental Panel on Climate Change (IPCC), take years to produce, years in which climate science continues to evolve.

Another issue is also at play: people. After all, volumes such as the periodical scientific assessments the IPCC makes (the most recent being *Climate Change 2001: The Scientific Basis* [hereafter *CC 2001*]) are compiled by individuals, each with their own emphasis, their own take, and, quite often, their own vested research interests. It is not surprising, for example, that *CC 2001* strongly emphasizes the controversial multicentury temperature history of Mann (1998, 1999), which largely lacks both the Little Ice Age and the Medieval Warm Period (and is known generally as the "hockey stick" chart, because of the shape of the curve used to describe it), given that Mann is one of eight lead authors of the chapter called "Observed Climate Variability and Change." In the bibliography to that chapter, Mann is the only *CC 2001* lead author who is listed as the principal author of any paper on millennial climate history. Several other long climate histories sketch quite a different shape from the now-famous hockey stick, and hundreds of individual papers present a result that is not nearly so stark as Mann's.

In chapter 2 of this volume, McKitrick comments on this problem, analogizing it to issues of corporate transparency:

> The failure of the IPCC to carry out . . . independent verification or to audit studies may be partly explained by the lack of independence between the chapter authors and the original authors. Professor Mann was lead author of the chapter relying on his own findings, a lack of independence that would never be tolerated in ordinary public offerings of securities.

Indeed, any compilation professing to represent "the state of the science" will necessarily reflect timeliness and author perspective and so will a book such as this one. Cohesiveness has its appeal, but it is multiple perspectives that allow for the exploration of all facets of a complex science. When addressing climate change, diversity of view should be welcomed and encouraged.

Here McKitrick describes the controversy that resulted from investigation of the Mann record. He concludes that "the story behind the Hockey Stick graph provides a cautionary tale about the need to understand the limited function of journal peer review, and the dangers of proceeding with major policy decisions, not to mention declarations of scientific consensus, without applying a further level of due diligence equivalent to an audit or an engineering study."

With *CC 2001*'s Summary for Policymakers (IPCC 2001) as its touchstone, this volume and the papers that comprise it exists to provide an expanded perspective on climate science today. Besides McKitrick's chapter, it includes a technical chapter on the nature of surface observations by Robert C. Balling; an updated analysis of upper-air temperatures, including satellite records by John Christy; an analysis of changes in extreme events and storms such as hurricanes and extratropical cyclones that goes beyond the TAR *(Third Assessment Report)* by Randall S. Cerveny; a summary of difficulties involved in the projection of precipitation changes by David R. Legates; a summary of new model findings on the relationship of El Niño and other large-scale circulation systems to overall climate variability by Oliver W. Frauenfeld; new studies on warming and mortality by Robert E. Davis; an expanded discussion of the influence of solar changes on climate by Sallie L. Baliunas; and an analysis of systematic problems with climate models by Eric S. Posmentier and Willie Soon.

But first, to the *CC 2001*'s Summary for Policymakers, and the misstatements, omissions, and misleading text contained therein. That summary is the place to start, since identifying its erroneous conclusions sets the tone for the exploration to come. Where appropriate, readers are referred to subsequent text in this volume that elaborates on the discussion here.

Incomplete Statement—and Misleading. Regarding global surface temperature records, *CC 2001* says, "These numbers take into account various adjustments, including urban heat island effects" (IPCC 2001, 2).

The reader can only conclude that the temperature records in *CC 2001* have been properly corrected for biases relating to urbanization and other "various adjustments," such as land-use changes. Since the publication of *CC 2001*, Kalnay and Cai (2003) found, using what was thought to be the very "clean" U.S. temperature history, that the biasing effect of land-use changes, which includes urbanization, was about three times larger than previously thought. In fact, applying their correction to the continental U.S. temperature history (figure 1.1) removes any significant warming effect from the record. It should be noted that this does not abrogate a global surface warming as the rise in temperature measured over the lower forty-eight states is generally accepted to be much lower than in many other parts of the world.

In chapter 3, Balling discusses corrections for urban heating within the IPCC context, concluding that "there is little question whether this warming has contaminated the near-surface temperature records that are used to

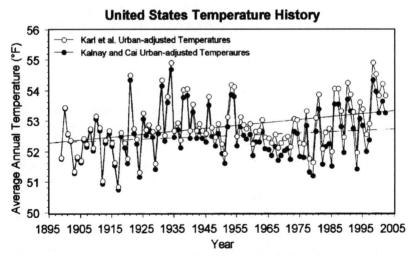

United States Temperature History

Figure 1.1.   U.S. temperature history from the National Climatic Data Center, adjusted for land-use effects in Kalnay and Cai (2003). This adjustment removes any statistically significant warming trend.

create the global temperature estimates." Balling also discusses the effects of desertification, as well as instrumental changes and adjustments, noting that "the fact that all biases are positive [in the warm direction] implies that the indicated [global temperature] rise of 0.7°C is nonetheless an overestimate."

Michaels et al. (2004) and McKitrick and Michaels (2004) have recently applied an entirely different methodology, specifically extracting "economic" signals from regional temperature trends and found pervasive warm bias as well as a greenhouse signal that was evident in cold dry regions in winter. They found that the spatial pattern of warming trends is significantly correlated with nonclimatic factors, including economic activity and sociopolitical indicators.

In chapter 3, Balling summarizes all that we know at this moment in time:

- Global near-surface air temperatures have risen in recent decades.
- Lower-troposphere [satellite-sensed] temperatures have warmed little over the same period.
- That differential is not consistent with predictions from numerical models.
- Proposed policy solutions will produce trivial results of virtually zero statistical significance.
- This era should be a time of scientific assessment, not a time of impulsive policy actions targeted at a problem that may not even be a problem and will not be significantly impacted by those very policy actions.

In addition, there are problems with spatial averaging techniques that can yield different global temperature trends. McKitrick notes that "different but equally valid averaging rules applied to the same data can yield conflicting results regarding 'the' global temperature, as has been seen recently with satellite data, for instance."

Misleading Statement. CC 2001 states that "globally it is very likely that the 1990s was the warmest decade, and 1998 the warmest year in the instrumental record, since 1861" (IPCC 2001, 2).

In that statement, the Summary for Policymakers fails to mention that the mid-nineteenth century is considered by virtually all scientists to be the end of a three-century cold period known as the Little Ice Age (LIA), a period that is anomalously absent in the "hockey stick" records that are emphasized in *CC 2001* (shown here as figure 1.2). The warming of the early twentieth century, which is as large as the late twentieth-century warming,

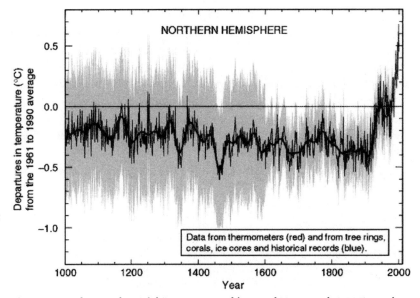

Figure 1.2. The "Hockey Stick" temperature history of Mann et al. (1999), as given in IPCC (2001).

is largely attributed to solar change by many scientists, including Lean and Rind (1998) and Rind (2002). Strong evidence also connects the length of the solar cycle to a significant portion of the late twentieth-century warming, as shown by Friis-Christensen and Lassen (1991) and Baliunas and Soon (1995). Consequently, it is not at all surprising, even without greenhouse effect changes, that the last years in the instrumental record should be among the warmest.

But rather than point out these widely known facts, the Summary for Policymakers further distorts the context.

Misleading Statement. "New analyses of proxy data for the Northern Hemisphere indicate that the increase in temperature in the twentieth century is likely to have been the largest of any century during the past 1,000 years. It is also likely that, in the Northern Hemisphere, the 1990s was the warmest decade and 1998 the warmest year" (IPCC 2001, 2).

That assumption is based on the findings of Mann et al. (1999). As shown by Soon et al. (2003) and Soon and Baliunas (2003), Mann's work appears to be an outlier compared with a vast body of scientific literature. It is the *only* "proxy" record of dozens that does not show a cold period from the seventeenth through the nineteenth centuries (the LIA) that accompanied the worldwide expansion of nonpolar glaciers, as well as a warm period near the beginning of the millennium, commonly called the Medieval Warm Period (MWP), whose warmth was similar to or even possibly greater than temperatures in the current regime. Soon et al. (2004) have detected a troubling "creep," in which the same records show increasing warming, that is attributable to Mann's analytical record. McIntyre and McKitrick et al. (2003) found several alterations and inconsistencies in the records used for Mann's analysis, and upon recalculation, produced a fifteenth century that contained an extensive warm period of higher temperature than the late twentieth century.

In chapter 2, McKitrick describes the discovery of problems with the Mann methodology and data. He concludes that the difficulties arose because of weak peer review:

> "Peer Review" for an academic journal is a much lesser form of due diligence than an audit of financial statements. The referees of our submission expressly stated that attempting to determine who was right or wrong between [Mann et al. 1998] and ourselves was far beyond the scope of review they could provide. Yet the differences are important to public policy.

Misleading Statement. "Since the late 1950s (the period of adequate observations from weather balloons), the overall global temperature increases in the lowest 8 kilometers of the atmosphere and in surface temperature have been similar at 0.1°C/decade" (IPCC 2001, 4).

The nonexpert reader would conclude that tropospheric temperatures measured from weather balloons and surface temperatures have risen in unison. On the contrary, climate models, in general, project a larger tropospheric warming than what is observed at the surface. As Christy shows in chapter 4, there is a profound disconnection between surface and upper atmospheric temperature trends. Christy is clearly troubled by the Summary for Policymakers. He writes:

> Why, for example, was a full page of the brief "Summary for Policymakers" devoted to surface temperature charts that depict considerable warming with another half-page of supporting text, while changes in the full bulk temperature of the atmosphere—far more important than the physics of the greenhouse effect—garnered but seven sentences?

Figure 1.3 shows the weather balloon and surface data concurrently for two periods. The first is from 1958 (the start of the balloon record) through 1976, and the second is the data since January 1, 1977. Clearly the second

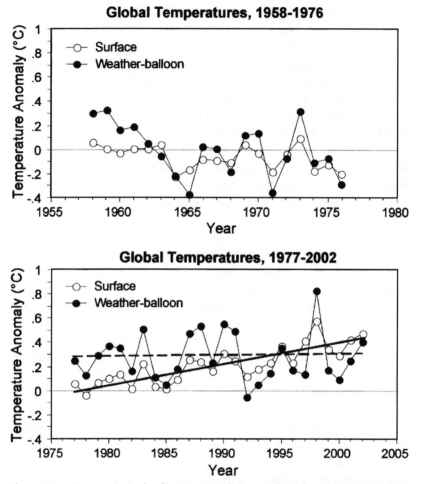

Figure 1.3. Temperatures in the 850–300mb (approximately 5,000–30,000 feet) layer show a step-change in the mid-1970s, contrary to projections of smooth changes by climate models for an enhanced greenhouse effect (data from Angell 2003).

period, which composes more than half the record, shows no significant warming in the weather balloon record, while the surface record shows a significant warming. The balloon data are the mean layer temperature from 850mb to 300mb, which is approximately 5,000 to 30,000 feet in elevation, depending upon temperature.

The only explanation for that is that there was a single step change in balloon-measured temperature that occurred in 1976. In fact, there is additional evidence for this in oceanic records, in a well-documented

1976–1977 phenomenon known as the "great pacific climate shift" (Miller et al. 1994). Yet there is no mention of that phenomenon in the Summary for Policymakers, and no citation of this well-known work in the subsequent chapter on observed climate change. That omission is discussed at greater length by Frauenfeld in chapter 7.

These facts disturb the facile conclusion that current warming is almost exclusively a greenhouse-related phenomenon, since no climate simulations produce such abrupt change under a greenhouse warming scenario. If we assume those models are correct, then it becomes difficult to ascribe observed warming of the troposphere to human influence. On the other hand, if we assume that this invalidates the models, then their future projections of temperature change, at least in the vertical dimension, are of proportionally less value.

Solar changes may also be a cause of some recent warming. In chapter 9, Baliunas discusses possible amplification of small solar perturbations by the earth's climate system, including alterations of the earth's magnetic field and ubiquitous ecosystem-scale responses through millennia. This persistent effect argues for some type of terrestrial magnification of what appears to be only small changes in solar output and effects on magnetic fields. But, she notes, if this has appeared for thousands of years, it must be continuing today.

Baliunas concludes, "While new empirical links between the sun and climate are being reported, their physical explanations lie in the realm of the poorly described, so cannot be ruled out on the scales of decades to millennia."

Incorrect Statement. In reference to the difference between the surface and balloon temperatures, the Summary for Policymakers states on page 4, "it is physically plausible to expect that over a short time period (e.g., 20 years) there may be differences in temperature trends," because of factors such as ozone depletion, atmospheric aerosols, and El Niño.

As figure 1.3 shows, the difference between the surface and the weather balloon temperatures is a result of the 1976 discontinuity, and it has therefore been in existence for twenty-eight years. Simply limiting the discontinuity to the concurrent period of MSU-satellite temperatures, 1979 to present, ignores the fact that it arose several years earlier. Again, the veracity of the satellite data are discussed extensively by Christy in this volume.

The National Academy of Sciences (2001) has commented on the disparity between surface and tropospheric temperatures:

> The finding that the surface and tropospheric temperature trends have been as different as observed over intervals as long as a decade or two is

difficult to reconcile with our current understanding of the processes that control the vertical distribution of temperature in the atmosphere.

It is obvious that the consensus of the National Academy on this major issue is at odds with the IPCC's.

Misleading Statement. "Satellite data show that there are very likely to have been decreases of about 10 percent in the extent of snow cover since the late 1960s" (IPCC 2001, 4). In general, as figure 1.4 shows, there is a strong correlation between hemispheric temperature and regional snow

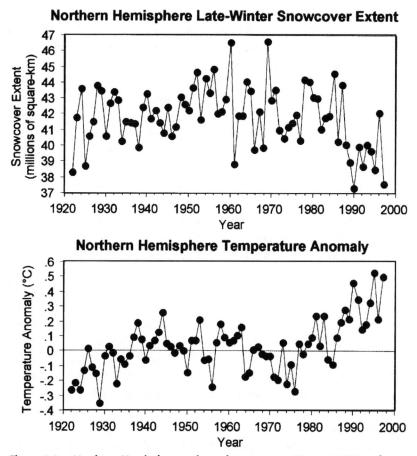

Figure 1.4. Northern Hemisphere estimated snow cover (Brown 2002) and temperature. Statements about changes since the late 1960s are misleading because they start out at a high point in the record.

cover. The late 1960s are at or near the coldest point for the period 1925–2000, so using that as a starting point is highly misleading.

Wrong and Misleading. On page 4, the Summary for Policymakers also states, "It is likely that there has been about a 40 percent decline in Arctic sea ice thickness during late summer to early autumn in recent decades."

This statement was *known to be wrong* at the time the TAR was published. It originates from analyses of submarine data by Rothrock et al. (1999) and a subsequent calculation by Wadhams and Davis (2000).

Other ongoing studies, which the IPCC had to be aware of, showed that sea ice was affected more by wind than by air temperature and that the areas of observed thickness decrease were compensated for by areas of increasing thickness that were not sampled by the submarines. Holloway and Sou (2002) demonstrated that the submarines were sampling the path where the ice was thinning fastest. So, by extrapolating this data to a model for the overall ocean, erroneously large estimates were made for thinning. According to Holloway and Sou, "When modeled thinning at the locations and dates [used in Rothrock's study] ranged from 25 percent to 43 percent, [our results for the entire Arctic] showed total thinning by lesser amounts ranging from 12 percent to 15 percent. . . . Even this lesser amount is quite specific to the timing of the observations." Further, they wrote, "If this is true then the [IPCC-] inferred rapid loss of ice volume was mistaken due to undersampling, an unlucky combination of ever-varying winds and readily shifting ice."

Misleading usage. Page 4 of the Summary for Policymakers states, "In the mid- and high latitudes of the Northern Hemisphere over the latter half of the twentieth century, it is likely that there has been a 2 percent to 4 percent increase in the frequency of heavy precipitation events."

That work originated largely in studies by Karl et al. (1995) and Karl and Knight (1998). The definition of heavy was originally 2 inches (50.8 cm) of rain per day. In reality, over the United States (and typical of much of the rest of the world), 70 percent of all rain days having more than 2 inches experience less than 3 inches. In the United States as a whole (the lower 48 states) there is no statistically significant increase in rains of 5 inches or more per day, an amount that would generally produce some flooding, while the statistically significant increase is confined to the Northeast (Michaels et al. 2002).

In chapter 6, Legates extensively addresses the quality of precipitation data and the disparities between model-projected and observed rain and snow. He demonstrates that misspecification of these important variables invalidates almost all other model projections, as surface and atmospheric moisture levels are central to determination of atmospheric temperature, pressure and wind fields.

Misleading statement. "Warm episodes of El Niño–Southern Oscillation (ENSO) phenomenon . . . have become more frequent, persistent, and intense since the mid-1970s, compared with the previous 100 years" (IPCC 2001, 5).

Grove (1998) determined El Niño strength for the past five hundred years using historical texts, mainly from southern Asia. He found that major El Niños abound, including ten "very severe" events in the past five hundred years, events as or more significant than those in the late 1990s. There is nothing in Grove's record that makes the most recent three decades appear particularly unusual. Further, by 2000, at the time that the TAR was being written, El Niño indices had returned to their norms for the past hundred years and have stayed in that range since.

Incomplete Statement. "The rate of increase of atmospheric CO_2 concentration has been about 1.5 parts per million (0.4 percent) per year over the past two decades" (IPCC 2001, 7).

Although that is technically true, it obscures the fact that, since the mid-1970s, the increase in concentration has been statistically indistinguishable between a low exponent (which is implied by a percentage) and a simple linear increase. An exponential increase is required to generate a constant rate of warming. If the increase is linear, the logarithmic response of temperature to CO_2 changes means that the observed linear warming will become slower as the thermal lag of the ocean catches up with the atmosphere. Assuming much of the lag is taken up by forty years, the observed CO_2 increases of the past quarter century, if truly linear, argue for a reduction in the rate of surface warming beginning around 2020.

Further, a statistical fit of emissions per capita data from the Energy Information Administration (figure 1.5) shows that, globally, emissions per capita reached a maximum in the mid 1980s. The statistical significance of that maximum (as opposed to a continued increase) is at the 1-in-10,000 level.

Misleading Statement. "Some recent models produce satisfactory simulations of current climate without the need for nonphysical adjustments of heat and water fluxes at the ocean–atmosphere interface used in earlier models. . . . Simulations that include estimates of natural and anthropogenic forcing produce the observed large-scale changes in surface temperature over the twentieth century" (IPCC 2001, 9).

Those two statements do not account for the arbitrary incoming radiation flux adjustment that occurs with the inclusion of "sulfate aerosols."

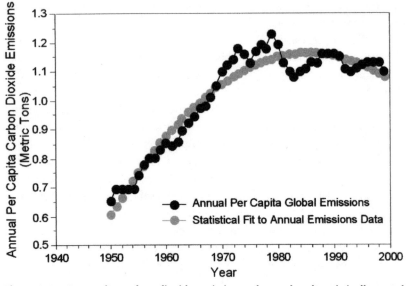

Figure 1.5.　Per capita carbon dioxide emissions, observed and statistically tested for a maximum. Data from the U.S. Energy Information Administration.

That term deserves quotation marks because, according to the IPCC, the "level of scientific understanding" of the radiative effect of aerosols is "very low." Figure 3 in the Summary for Policymakers gives a range of forcing of from 0 to -2 W/m^2, which implies a temperature effect of between 0 and 1.5°C.

The Summary for Policymakers's figure 4c shows a remarkable agreement between model output and observed global temperatures. Most striking is the similarity of the period between 1940 and 1975, when temperatures decline slightly. That period (especially the early part of the interval) has relatively large "sulfate" changes compared with greenhouse gases. Given that any level of forcing—from 0 to -2 W/m^2—can be selected, it is obvious that any model can be tuned to reproduce this seemingly troublesome interval, and they obviously have been.

In chapter 10, Posmentier and Soon analyze the internal dynamics of climate models. They argue that a science-based policy must be able to confidently calculate the trajectory of future CO_2 concentrations, the climatic effects of that trajectory, and the differences between climate change forced by that trajectory and change which arises as a result of natural variability on the regional scale.

They conclude, "Our current lack of understanding of the earth's climate system does not allow us to determine reliably the magnitude of the climate change that will be caused by anthropogenic CO_2 emissions, let alone whether this change will be for better or for worse."

That statement is not inconsistent with the finding of Michaels and Balling (2000) and Hansen and Sato (2001) (see below), that the warming we will see in the next fifty years is known to a very small range of error. That is because the trajectory for carbon dioxide is less relevant on that time scale, as much of the climate response is indeed integrating past changes. In general, Posmentier and Soon are concerned about a longer time horizon.

They further conclude that "systematic problems in our inability to simulate present-day climate change . . . paint a dismal picture of the difficult task ahead." Even with perfect knowledge of future CO_2 levels, they conclude that other variables inherent in the climate system have yet to be sufficiently defined. Unlike the Summary for Policymakers, they conclude that "climate specialists should continue to urge caution in interpreting GCM [climate model] results and to acknowledge the incomplete state of our current understanding of climate change."

Ironically, Posmentier and Soon close with a quotation from University of Wisconsin climatologist Reid Bryson, who is arguably the scientist most responsible for the modern notion that human beings can significantly change global and regional climate. A partial quotation of that is relevant to IPCC's contention:

> A model is *nothing more* than a formal statement about how the modeler believes the part of the world of his concern actually works . . . it may be years before computer capacity and *human knowledge* are adequate for reasonable simulation . . . the main models in use all have similar errors, but it is hardly surprising, for they are all essentially clones of each other.

Incomplete Statement. "Reconstructions of climate data for the past one thousand years also indicate that [the warming of the past one hundred years] was unusual" (IPCC 2001, 10).

Why does the IPCC insist on emphasizing only the climate reconstructions of Mann et al. (1998, 1999), which are largely devoid of well-known features such as the Little Ice Age (LIA) or the Medieval Warm Period (MWP)? The IPCC was surely aware of ongoing studies subsequently published by Esper et al. (2004) indicating much more variability in climate history. Further, as shown by Soon et al. (2003), who merely summarized a scientific literature that was largely in existence at the time the TAR was written, the balance of evidence strongly suggests that both the LIA and the MWP were real, and it is certainly a reasonable argument that the MWP was as warm as much of the twentieth century. Further, McIntyre and

McKitrick (2003) have shown that Mann's results are highly sensitive to data selection and analytical methods, and that they can easily yield warmer periods than the present with slightly different input.

McKitrick discusses these issues in chapter 2. He demonstrates that the data selection and analysis techniques used result in an overemphasis of climate change in recent years. In one remarkable graphic, McKitrick shows how Mann et al. (1998) were aware of that problem and actually ran an analysis that eliminated it, but declined to submit it for publication.

Important Implications Not Noted. "On the timescales of a few decades, the current observed rate of warming can be used to constrain the projected response to a given emissions scenario despite uncertainty in climate sensitivity. This approach suggests that anthropogenic warming is likely to lie in the range of 0.1°C to 0.2°C per decade over the next few decades" (IPCC 2001, 13).

Here the IPCC acknowledges the argument first made by Michaels and Balling (2000), and subsequently by Allen et al. (2000), and Hansen and Sato (2001), that future warming within the policy-relevant time frame is known to a rather small range of error and that it is at the lower limit of the range the IPCC gives. Hansen and Sato (2001) state that "future global warming can be predicted much more accurately than is generally realized . . . a warming of 0.75°C by 2050," and later, "we predict an additional warming in the next 50 years of 0.75°C +/− 0.5°C, a warming rate of 0.15 +/− 0.05°C per decade."

Wrong. Table 1, page 15, of the Summary for Policymakers states that the "confidence in observed changes" in the last half of the twentieth century concerning "increases in tropical cyclone peak wind intensities" is "not observed in the few analyses available."

In fact, as shown by Cerveny in chapter 5, Wilson (1999) examined strong hurricanes in the Atlantic Basin from 1950 through 1998 and found a *decreasing* trend and fewer intense storms during the warmer periods within that interval. Landsea et al. (1996) found a statistically significant *decrease* in average maximum winds measured by hurricane hunter aircraft since they began flying after World War II. That is radically different from the notion the IPCC puts forth, which is simply that there is little confidence that an "increase" in peak wind has been observed.

Cerveny notes an uptick in severe hurricane activity in the last five years of the twentieth century, but that is obviously too short a period for any causal implication. On the overall subject of hurricanes, Cerveny quotes Raghavan and Rajesh (2003): "Contrary to the common perception that tropical cyclones are on the increase, due perhaps to global warming, studies all over the world show that, although there are decadal variations, there is no definite long-term trend in the frequency or intensity of tropical cyclones over the period of about a century for which data are available." That sentence, taken from an extensive literature review, is sharply different from IPCC's statement on page 15. Clearly, the IPCC does not reflect the consensus of hurricane scientists here.

On the general subject of increased storminess, Cerveny writes that "an enormous disconnection exists between popular perception and scientific reality regarding the effects of global climate change on severe weather and natural disasters." He concludes, "Fundamentally, the weight of evidence, regarding the impact of increasing atmospheric carbon dioxide and global warming on severe weather, suggests . . . climate change impact, at the present time, is small, if present at all, and very difficult to adequately identify."

Addressed in this Volume. Page 16 of the Summary for Policymakers states, "Even with little or no change in El Niño amplitude, global warming is likely to lead to greater extremes of drying and heavy rainfall and increase the risk of droughts and floods that occur with El Niño events in many different regions."

That conclusion is based on the notion, offered later in the *CC 2001*, that the teleconnections between temperature and precipitation anomalies and El Niño "may shift somewhat" in a warmer world. Frauenfeld (chapter 7) closely examines models for El Niño and finds them largely inadequate. Therefore projections of changes in teleconnections are simply unreliable.

With regard to El Niño, *CC 2001* states, "Analyses of several global climate models indicate that as temperatures increase due to increased greenhouse gases, the Pacific climate will tend to resemble a more El Niño-like state. . . . A majority of models show a mean El Niño-like response in the tropical Pacific." In addition, *CC 2001* is confident about teleconnections between El Niño intensity and other climate anomalies, saying model results:

> Indicate that future seasonal precipitation extremes associated with a given ENSO event are likely to be more extreme due to the warmer, more El Niño-like mean base state in a future climate . . . also in association with changes in the extratropical base state in a future warmer climate, the teleconnections may shift somewhat with an associated shift of precipitation and drought conditions in future ENSO events.

Frauenfeld demonstrates that models are indeed quite unreliable in simulating observed tropical variability in the Pacific, and that therefore simulations of the extratropical (i.e., temperate and polar) responses to tropical forcing must be inadequate. This results from fundamental misspecification of the vertical profile of tropical temperature that is inherent in climate models, as discussed here by Christy and Posmentier and Soon.

Inadequate Basis. "It is likely that warming associated with increasing greenhouse gas concentrations will cause an increase of Asian summer monsoon precipitation variability" (IPCC 2001, 15).

That is an easily testable hypothesis, as the IPCC states on page 10 that "detection and attribution studies consistently find evidence for an anthropogenic signal in the climate record of the last thirty-five to fifty years." Consequently, we can test whether there is an increase in the interannual variance of monsoon rainfall along with anthropogenic warming. Section 2.6.4 (Observed Climate Variability and Change: Monsoons in *CC 2001*) contains no reference indicating detection of increased variability despite the contention of a half century of anthropogenic warming.

Inadequate Basis. "Beyond 2100, the thermohaline circulation could completely, and possibly irreversibly, shut down" (IPCC 2001, 16).

Carl Wunsch, who is largely recognized as the world's authority on ocean currents, recently commented on that possibility: "The only way to produce an ocean circulation without the Gulf Stream is either to turn off the wind system, or to stop the earth's rotation, or both" (Wunsch 2004).

Misleading Statement. "After greenhouse gas concentrations have stabilized, global average surface temperatures would rise at a rate of only a few tenths of a degree per century rather than several degrees per century as projected for the 21st century" (IPCC 2001, 17).

A growing number of scientists now places warming for the next fifty years at around 0.75°C (recall comments about the Summary for Policymakers, page 13, earlier in this chapter). Further, nearly all models show largely linear rather than strongly exponential warming on the century scale. Figure 1.6, taken from figure 9.3 of the *CC 2001*, depicts that behavior. Notably, all the models shown are driven with a 1 percent per year increase in carbon dioxide while the actual rate of increase has been, as noted above, statistically indistinguishable between simple linear increase, and a much smaller exponent of 0.4 percent per year over the course of the past quarter century.

The model response of carbon dioxide is highly linear with respect to concentration. Hansen and Sato (2001) suggests that the net annual increase in greenhouse forcing, in carbon dioxide equivalents, has settled at around 0.7 percent per year. Consequently, the consensus of models given in figure 1.6 (*CC 2001,* figure 9.3) should be adjusted downward, from roughly 2.6°C by 2100 to 1.8°C, which is hardly "several degrees per century."

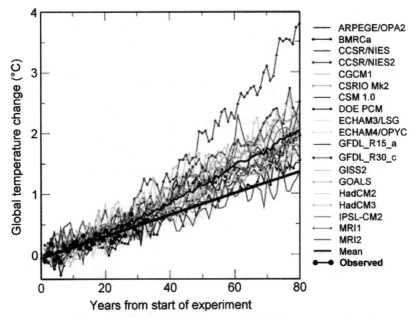

Figure 1.6. The central tendency of climate models is for a constant (not increasing) rate of warming, as has been observed (dots and line). Figure modified from IPCC 2001.

This brief summary examines but a few misleading or incomplete statements from the section in the IPCC's *Climate Change 2001: The Scientific Basis* characterized as a "summary for policy makers," a section designed to be read most often by the most people—whether policymakers, journalists, scientists, or private citizens. Its influence is therefore significant, and its misstatements and errors egregious. Though such a summary can and should be able to stand alone, a more thorough understanding of its conclusions is available within the chapters that follow. Like *CC 2001*'s summary, this introduction extracts the main points of the chapters to come. But ultimately it is designed to serve as a guide to the content of this volume, where detailed analyses of climate history, models, and the correspondence between the two reveals a world much different than that portrayed in *CC 2001*.

REFERENCES

Allen, M. R., P. A. Stott, J. F. B. Mitchell, R. Schnur, and T. L. Delworth. 2000. Quantifying the uncertainty in forecasts of anthropogenic climate change. *Nature* 407:617–20.

Angell, J. K. 2003. Global, hemispheric, and zonal temperature deviations derived from radiosonde records. In *Trends Online: A Compendium of Data on Global Change*. Carbon Dioxide Information Analysis Center, Oak Ridge National Laboratory, U.S. Department of Energy, Oak Ridge, Tenn.

Baliunas, S., and W. Soon. 1995. Are variations in the length of the activity cycle related to changes in brightness in solar-type stars? *Astrophys Journ* 450:896–901.

Brown, R. 2002. Reconstructed North American, Eurasian, and Northern Hemispheric snow extent, 1915–1997. National Snow and Ice Data Center, Boulder, Colo.

Esper, J., D. C. Frank, and R. J. S. Wilson. 2004. Climate reconstructions: Low-frequency ambition and high-frequency ratification. *EOS* 85:113, 120.

Friis-Christensen, E., and L. Lassen. 1991. Length of the solar cycle: An indicator of solar activity closely associated with climate. *Science* 254:698–700.

Grove, R. H. 1998. Global impact of the 1789–93 El Niño. *Nature* 393:318–19.

Hansen, J. E., and M. Sato. 2001. Trends of measured climate forcing agents. *Proc Natl Acad Sci* 98:14778–83.

Holloway, G., and T. Sou. 2002. Has Arctic sea ice rapidly thinned? *J Climate* 15:1691–1701.

Intergovernmental Panel on Climate Change (IPCC). 2001. *Climate change 2001: The scientific basis*. Cambridge: Cambridge University Press.

Kalnay, E., and M. Cai. 2003. Impact of urbanization and land use change on climate. *Nature* 423:528–31.

Karl, T. R., and R. W. Knight. 1998. Secular trends of precipitation amount, frequency, and intensity in the USA. *Bull Amer Meteor Soc* 79:231–41.

Karl, T. R., R. W. Knight, and N. Plummer. 1995. Trends in high-frequency climate variability in the twentieth century. *Nature* 377:217–20.

Landsea, C. W., N. Nicholls, and L. A. Avila. 1996. Downward trends in the frequency of intense Atlantic hurricanes during the past five decades. *Geophys Res Lett* 23:1697–706.

Lean, J., and D. Rind. 1998. Climate forcing by changing solar radiation. *J Climate* 11:3069–94.

Mann, M. E., R. S. Bradley, and M. K. Hughes. 1998. Global-scale temperature patterns and climate forcing over the past six centuries. *Nature* 392:779–87.

———. 1999. Northern Hemisphere temperatures during the past millennium: Inferences, uncertainties, and limitations. *Geophys Res Lett* 26:759–62.

McIntyre, S., and R. McKitrick. 2003. Corrections to the Mann et al. 1998 proxy data base and Northern Hemispheric average temperature series. *Energy and Environment* 14:751–71.

McKitrick, R., and P. J. Michaels. 2004. A test of corrections for extraneous signals in gridded surface temperature data. *Clim Res* 26:159–73.

Michaels, P. J., and R. C. Balling Jr. 2000. *The satanic gases*. Washington, D.C.: Cato Institute Press.

Michaels, P. J., P. C. Knappenberger, R. E. Davis, and O. W. Frauenfeld. 2002. Rational analysis of trends in extreme temperature and precipitation. Presented at the Thirteenth Conference on Applied Climatology, Portland, Ore.

Michaels, P. J., R. McKitrick, and P. C. Knappenberger. 2004. Economic signals in global temperature histories. Presented at the American Meteorological Society annual meeting, Seattle.

Miller, A. J., D. R. Cayan, T. P. Barnett, N. E. Graham, and J. M. Oberhuber. 1994. The 1976–77 climate shift of the Pacific Ocean. *Oceanogr* 7:21–26.

National Academy of Sciences. 2001. *Climate change science: An analysis of key questions.* Washington, D.C.: National Academy Press.

Raghavan, S., and S. Rajesh. 2003. Trends in tropical cyclone impact: A study in Andhra Pradesh, India. *Bulletin of the American Meteorological Society* 84:635–37.

Rind, D. 2002. The sun's role in climate variations. *Science* 296:673–77.

Rothrock, D. A., Y. Yu, and G. A. Maykut. 1999. Thinning of the Arctic sea-ice cover. *Geophys Res Lett* 26:3469–73.

Soon, W., and S. Baliunas. 2003. Proxy climatic and environmental changes of the past 1,000 years. *Clim Res* 23:89–110.

Soon, W., S. L. Baliunas, C. Idso, S. Idso, and D. R. Legates. 2003. Reconstructing climatic and environmental changes of the past 1,000 years: A reappraisal. *Energy and Environment* 14:233–96.

Soon, W., D. R. Legates, and S. L. Baliunas. 2004. Estimation and representation of long-term (>40 year) trends of Northern-Hemisphere-gridded surface temperature: A note of caution. *Geophys Res Lett* 31, doi:10.1029/2003GL019141.

Spencer, R. W., and J. R. Christy. 1990. Precise monitoring of global temperature trends from satellites. *Science* 247:1558–62.

Wadhams, P., and N. R. Davis. 2000. Further evidence of ice thinning in the Arctic Ocean. *Geophys Res Lett* 27:3973–75.

Wilson, R. M. 1999. Statistical aspects of major (intense) hurricanes in the Atlantic basin during the past 49 hurricane seasons (1950–1998): Implications for the current season. *Geophys Res Lett* 26:2957–60.

Wunsch, C. 2004. Gulf Stream safe if wind blows. *Nature* 428:601.

2

THE MANN ET AL. NORTHERN HEMISPHERE "HOCKEY STICK" CLIMATE INDEX

A Tale of Due Diligence

Ross McKitrick

This chapter tells the story of the detective work of Stephen McIntyre (and, to a lesser extent, myself) regarding the famous "hockey-stick" climate history graph of Mann, Bradley, and Hughes (1998), better known as MBH98.[1] After studying in detail how the hockey-stick graph was done, we found mistakes in the data and methods that went unnoticed for years, even as the graph was used by governments worldwide to drive major policy decisions. The story behind the hockey stick provides a cautionary tale about the need to recognize the limited function of journal peer review and the dangers of proceeding with major policy decisions without applying a further level of due diligence equivalent to an audit or an engineering study. It also shows that the Intergovernmental Panel on Climate Change failed to carry out elementary due diligence on its most famous promotional graphic, despite widespread perceptions that it had.

In 1998, Michael Mann, Raymond Bradley, and Malcolm Hughes (hereafter MB&H) published a paper in *Nature* called "Global-scale temperature patterns and climate forcings over the past six centuries," commonly called MBH98. In this study, they proposed some seemingly novel ways of calculating the average Northern Hemisphere temperature back to AD1400. In 1999, MB&H published a follow-up paper, which is commonly called MBH99, extending MBH98 back four hundred years to AD1000.[2] MBH99 did not recalculate post-1400 values; it simply extended the previous results to an earlier period. This is Mann's famous hockey-stick curve (figure 2.1).

This graph achieved notoriety courtesy of the Intergovernmental Panel on Climate Change (IPCC), appearing in figures 2-20 and 2-21 in

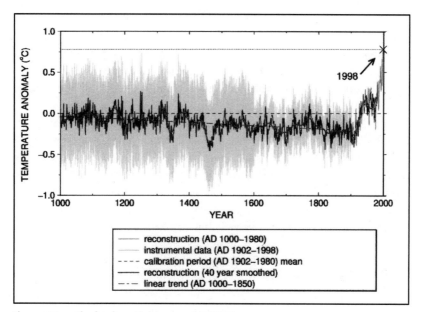

Figure 2.1.   **The hockey stick curve of MBH99.**

chapter 2 of the 2001 Working Group 1 Assessment Report, figure 1b in the Working Group 1 Summary for Policymakers, figure 5 in the Technical Summary, and figures 2-3 and 9-1B in the Synthesis Report. The IPCC Summary for Policymakers (3) used this figure to claim that it is likely "that the 1990s has been the warmest decade and 1998 the warmest year of the millennium" for the Northern Hemisphere. The hockey-stick graph has been reprinted countless times and used by governments around the world as the official, canonical climate history of the world.

MBH98 was conspicuous for its obscurity. The appendix to that paper, which spells out the statistical procedures, is very hard to read. It is written in grandiose yet disorganized prose and omits the mathematical equations that would allow expert readers to attain an unambiguous understanding of what was done. For all the subsequent usage of the results of that paper, it strikes me as conspicuous that the methods of the paper have not been widely applied. Even Mann himself did not use its method in his recent publications.[3] My suspicion is that most readers cannot make heads or tails of Mann's methodology. Nevertheless, the hockey-stick graph was an instant hit. Apparently no one ever checked it, and yet it was used by institutions, governments, and the media to promote the Kyoto Protocol.

THE AUDIT: APRIL TO SEPTEMBER 2003

Hockey-stick graphs are notorious promotional tools for people in business and finance: dot-com promotions typically have hockey-stick shapes to describe future revenue and sales projections. After the dot-com boom, however, many businesspeople cringe when they see a hockey-stick graph.

In late 2002, the Canadian government was trying to sell the Kyoto Protocol to the public and was continually citing Mann's hockey stick as one of its central arguments for supporting the costly investments needed to implement Kyoto. At the time, Stephen McIntyre was a fifty-five-year-old Canadian businessman involved in financing speculative mineral exploration. In that business, McIntyre knew the importance of effective promotional graphics in raising money, but he also knew enough to be wary of them. He had dealt regularly with geologists, one of whom had pointed out to him that the climate change in recent times was far less than that observed in geological history. When Stephen looked at Mann's hockey stick, the graph immediately struck him as a promotional graphic of the type that he was used to seeing in business. At the time, the speculative mining business was slow, and he thought it would be interesting to see how this particular promotion was designed.

Note that this impulse was no more than the intuitive suspicions of someone with long experience in the speculative mineral exploration business. That was the start of the whole adventure. By no means did McIntyre begin his inquiry as a hatchet job on Mann and his coauthors. On the contrary, it never occurred to him that anything would result from his efforts other than figuring out a few things of personal interest. Nor was this an exercise in pouncing on small errors. The resulting critique centers on the lack of transparency in the original work, which is what made it so difficult to find the errors therein, and the overhasty promotion of the graph by governments and others who were making no attempt to check the calculations. That the errors we found were material to the results only highlights the need for greater transparency at the journal level, and more rigorous due diligence by users of the research, points to which I will return in the concluding section.

One thing Stephen McIntyre learned from the mining business was that it is important to look at the drill core—you need to see the raw data, not just the promoter's version. In the famous Bre-X fraud of 1998, the promoters said they had a "special" analyzing technique that required the entire core—so that visiting analysts were unable to inspect the original core. In this case, Stephen concluded that the equivalent to the "drill core" would be the individual proxy series. He decided to plot up these data to see what they looked like.

Although the long IPCC graph came from MBH99, the methods were developed in MBH98, so the results could not be analyzed without looking at the earlier study. After searching the relevant FTP and websites, Steve was able to locate the data for fourteen series in MBH99, but not for MBH98. So on April 8, 2003, on a whim, he wrote to Michael Mann and asked him where the data could be obtained, not expecting any special effort on his behalf, merely presuming that the data had been neatly buttoned up for previous due diligence by the IPCC and others. Mann replied that it was on an FTP site they had set up. Mann added, "I've forgotten the exact location, but I've asked my colleague Dr. Scott Rutherford if he can provide you with that information." A few days later Steve queried Rutherford, who responded that the data "aren't actually all in one FTP site, at least not to my knowledge. I can get them together if you give me a few days." Eleven days later, Steve received the FTP location of a text file, pcproxy.txt, which, after downloading, was found to contain 112 columns of data, matching the number of proxies identified in the *Nature* paper.

The episode immediately caught his attention. He had assumed that the IPCC and others had carried out due diligence prior to relying on this graphic and that such due diligence would have necessarily required examination of the data—drawing from his own experience with audits and business due diligence. If the data had not been so organized, was it possible that no one had ever checked the data? It was a bizarre possibility, but the collective failure of due diligence in the Bre-X (and, for that matter, Enron) collapses were just as strange. People in the mineral exploration business are used to strange events. And someone accustomed to public markets knows about the "madness of crowds."

So instead of simply trying to plot up the proxy data and see what it looked like, Steve decided to do something much more ambitious. He decided to try to verify MBH98 by replicating its calculations. In later months I was sometimes asked for a brief description of my coauthor, and inevitably the people doing the asking were puzzled about the answer: a middle-aged mining businessman. What was a mining executive with no specialized training doing walking into a high-level scientific debate with international policy implications? However the situation was remarkably propitious. As far as credentials go, Steve had been a prize-winning student in math and statistics at the University of Toronto and had won a Ph.D. scholarship offer from MIT. But for a decision to go into business he could easily have had a distinguished academic career. Even though he hadn't done any math for nearly thirty-five years, he soon found the math in MBH98 wasn't very difficult once you got past the convoluted description. Steve figured it would be an interesting exercise and was willing to put the time into it. What a lucky break for science.

MANN MINUTIAE AND MORE MANN MINUTIAE

After collecting the source data, Steve began looking at the individual proxies, the way that a geologist would look at drill core, and began making a series of postings on the internet group "climatesceptics" under the heading "Mann minutiae," followed by "More Mann minutiae," telling people about his work. His comments caught the interest of, among others, Dave Douglass of the University of Rochester and Bob Carter of James Cook University in Australia, who encouraged him in the specific analysis of the individual proxies, saying that this work was worthwhile and no one else was doing it.

This he began to do in earnest in the summer of 2003. Watchers of the climate skeptics list would see periodic postings from him, but none of us knew anything about him except that he seemed to be doing a lot of fascinating work on MBH98. In late July, Steve contacted me to say that he lived in Toronto (the University of Guelph is nearby) and suggested we get together to talk about this project. We finally did so in September, meeting at almost the exact moment when Hurricane Isabel hit Toronto.

Sonja Boehmer-Christiansen, the editor of *Energy and Environment,* had been interested in Steve's postings and asked him to consider submitting a paper. He agreed, flattered that his notes had occasioned any interest.

PRINCIPAL COMPONENTS

With that encouragement, he decided to try to replicate the principal components calculations in MBH98. "Principal components" analysis is a statistical method of reducing the dimension of a large data set. A principal component series (PC) is essentially a weighted average of the original group of series, with the weights chosen in a special way to maximize "explained variance." A data set has an associated sequence of PCs (denoted PC1, PC2, etc.), each one accounting for successively less and less of the variance of the underlying data. The algorithm is included in modern statistical languages and requires a single line in R (which is the language that Steve was using) to execute.

MBH98 used PC analyses to simplify both temperature and proxy data. For temperatures, they simplify 1082 grid cell series to sixteen principal component series. For proxies, MBH98 describes 112 proxies, of which 71 are individual records and 31 are PCs computed from six "networks" of individual proxies, containing more than three hundred original series. The networks are from geographical regions with labels like "NOAMER" (North America) and "SWM" (Southwest–Mexico). The maximum number of retained PCs for each region varies from three to nine. Details of these calculations were never published by Mann, but the sites for five of six regions were listed on a Supplementary Information (SI) page at *Nature*. Aside from upgrading his computer savvy and refreshing his math skills, Steve had a formidable task in merely collecting, collating, and verifying the original data from the World Data Center for Paleoclimatology for the listed sites, which often had slightly varying formats.

He first tried to replicate Mann's temperature PC calculations using the most recent IPCC data. PC calculations only work when there are no missing data. In this case, he found that there were large amounts of missing data and the PC calculations wouldn't go through. In early September 2003, he e-mailed Mann to ask how he had dealt with missing data. He received no reply.

After laboriously collecting and verifying the tree-ring data, he calculated the PCs for five networks and compared the results to those which Mann had provided. They didn't match at all. Naturally he assumed he had erred in his own calculations and began to cross-check each step of the calculations.

He had used a standard algorithm in a high-level programming language (R), so he couldn't see that an error could have occurred at this stage. He calculated the explained variances from his PCs and those using the PC series from Mann. If a PC has been computed properly, the first PC should explain the most variance in the underlying data matrix. But for some networks, Steve obtained explained variance levels as low as 6 percent from the Mann data—which was impossible. After a while, he wondered whether he had somehow collated the data incorrectly when he read the data into R. This led him to examine the data set visually, whereupon he saw that many of the PC series that he used began in rows where the year ending is *99 or *49, while Mann's usual practice was to start series in years ending in *00 or *50. On a hunch, he shifted one of the PC networks up a year, and the explained variance shot upward. Further inspection confirmed that this was a problem in the original file, not in his own collations. He emailed Scott Rutherford about the problem, but Rutherford said that the data were from before his time and that Steve would have to contact Mann himself.

Steve also checked what happened at the end of the series. It turned out that the 1980 values for many PC series were identical to seven decimal places. It appeared that there had been gross collation errors in MBH98. Ironically, we would later learn that this problem was in the particular file (pcproxy.txt) Steve had obtained back in April, but likely did not affect MBH98 itself. Still, it was this error that indicated there were significant flaws in the database, leading us to undertake the work that uncovered the errors that did drive MBH98 results.

By this time, Steve was confident in his own PC series for each of the five regions where sites had been identified. In each case, his calculations resulted in very different series from those that Mann had obtained. The explained variance for his own calculations easily surpassed the explained variance from the PC series provided to us. We couldn't yet figure out what was wrong with the MBH98 calculations, but it was evident that something was.

While Steve was looking at the 1980 values, he noticed that the closing values for other PC series were simply "fills": extrapolations from an earlier year. Steve had just noticed these points when we met. Prior to our lunch, Steve had emailed a long paper he had drafted discussing his work up to that point. It was interesting, but the discussion was unfocused and the conclusions unclear. I agreed to work with him to help sharpen the focus and edit the paper down to a publishable form. In our collaboration since then, Steve has been the one doing all the hard stuff with data and analysis, while my role has been to ask questions, read his work, edit/rewrite, and help plan the research strategy. It's been a good collaboration.

At lunch, to avoid any possible misunderstandings, we agreed that it was time to send the entire data set to Mann and ask him for confirmation that this was the data actually used in MBH98. Steve did this. In reply, Mann said that he was too busy to respond to this or any further inquiries and referred us to a publication by Zorita et al., who, he said, had had no trouble in replicating his methods. In the aftermath of our later E&E publication, we took some flak for not giving MB&H more opportunity ahead of time to address our concerns before our paper was published. But the fact is that Mann had categorically terminated our correspondence without answering reasonable questions about his methodology and the provenance of the data set. With a clear message that Mann did not want Steve to contact him any further, we could hardly keep sending our analysis to him; I doubt he would have responded if we had. Indeed later, Mann continued to refuse to provide information on his data and methods.

Meanwhile, I asked Steve to send me the data. I carried out my own visual inspection of the data, finding new examples of filled-in values including one pair of series where the two different columns contained identical data for nearly thirty years. I suggested that we focus first simply on the implications of the data problems, leaving methodological discussions for another day.

That discovery persuaded Steve to tear apart the whole data set and look at every series from scratch. He began trying to identify original source data for every series in MBH98 and comparing the original data to the version used in MBH98, finding that the versions used in MBH98 were frequently different from publicly archived versions. The differences were generally due to the apparent use of obsolete versions in MBH98.

In other cases, he found that MBH had truncated series without any explanation and that some of the series listed in the Supplementary Information were not actually used. While this was going on, Steve had also been patiently trying to decode the obscure methodological descriptions of MBH98, turning them into linear algebra (which he remembered from his youth) and trying to see if he could replicate MBH98 results. Using the data provided to him, and following the published methods in MBH98 as closely as possible, Steve could get a hockey stick–shaped graph that was close to the original results, but not an exact replication. However, the emulation was clearly good enough to be used to study the effect of the various data problems that we had identified.

The $64,000 question was whether the data problems affected any results. Steve plugged in the new data set, with freshly calculated PCs and up-to-date data where updated versions had been identified, with no expectation of what would result. The result was the graph that we soon published, showing that early fifteenth-century values exceeded twentieth-century values, contradicting the MBH98 conclusion of twentieth-century uniqueness. Rumors of this finding began to spread quickly.

The conclusion of our paper was that "the extent of errors and defects in the MBH98 data means that the indexes computed from it are unreliable and cannot be used for comparisons between the current climate and that of past centuries." Notice I am not claiming we rediscovered a Medieval Warm Period. We have tried to be careful not to suggest we're proving the fifteenth century was "warm" compared with today. We simply argued that this was the result of applying MBH98 procedures to updated and properly collated data.

We wrote up these findings and, knowing they were provocative, we sent the draft to many readers, which resulted in a lot of constructive criticism. We also engaged a professional statistician who consults to paleoclimatology labs to write a report for us evaluating our paper. When we were satisfied that it was our best shot, we sent it to *Environment and Energy*. The referees asked for some changes, which we made, and the paper was accepted quickly. Anticipating the impact it would have, Bill Hughes, the publisher of *Environment and Energy* kindly waived the copyrights and allowed the paper to be posted on the MultiScience website for free distribution at the end of October, in a paper now known as "MM03."[4]

In our article, in addition to describing our findings, we also tried to document clearly our methodological decisions on matters where MBH98 procedures remained unreported, and provided Supplementary Information on a website (www.climate2003.com) containing the computer code used for our calculations.

One such decision related to the vexing problem of missing data in PC calculations, which was not described in MBH98 and which Mann had failed to elucidate upon inquiry. In Steve's program, he calculated tree-ring PC series over the maximum period for which all sites in the network were available. In some cases, that led to longer series than used in MBH98 and in some cases it led to shorter series, which we noted in our article. At this stage, we knew that there was obviously something wrong with the MBH98 PC series on multiple counts: the odd start dates, the identical 1980 values, and the low explained variance. We did not know precisely what was wrong with the MBH98 PC series. The difference in series lengths seemed merely one more strange defect. On the other hand, our own calculations were done with fresh data and could be verified as having a higher explained variance on the underlying data than those used by MBH98, so we were on solid ground with using these PCs, even though we did not yet know why the MBH98 proxy PCs were coming out wrong.

In the aftermath of publication, the resulting publicity forced Mann to disclose much previously unavailable information, including the location of a vast amount of FTP data unavailable to MM03. The new information has allowed us to nail down exactly how Mann's hockey-stick curve was constructed. The answer does indeed lie in the PC series and our *E&E* conclusions have been vindicated on grounds we had not anticipated at the time.

MM03 AND RESPONSES

MM03 attracted interest right away. Comments and news articles appeared in United States, Canada, Australia, Holland, Norway, Germany, Argentina, and several other countries. We attracted a great deal of attention in specialized blogs and chat lines. Since the U.S. Senate was debating the McCain–Lieberman bill, there was considerable interest in our work in Washington. Steve and I did a briefing on Capitol Hill on November 17, 2003. A few partisans like Mike McCracken and Stephen Schneider editorialized against us, but by and large the "mainstream" response was to acknowledge that we had made a legitimate critique of the hockey-stick results, that if we were right it would have serious repercussions in the global warming debate, and that there would have to be some detailed discussions in the months ahead to settle the matter fully.

Michael Mann's first reaction to our paper was in the form of some comments provided to a U.S. website, www.davidappell.com. For one thing, Mann insisted that we had analyzed the wrong data set. The file we looked at (he said) was a special-purpose collation put together in Microsoft Excel in response to Steve's inquiry back in April, as a "courtesy," since Steve was supposedly unwilling to get the data from Mann's FTP site. Supposedly some mistakes were inserted into the Excel file, and the resulting data set bore no resemblance to the one used for MBH98. He opined that we ought to have gone to his FTP site, where the real data are, and that had we done so we would have discovered the data we were using were flawed and we would then not have produced the results we did. Mann also gave out a new and different URL for the location of the MBH98 data, a URL that was not referred to at Mann's own website or in any other public source to that date.[5]

Mann's statements were untrue. Steve had not originally asked for a data file but for an FTP site. Mann had told him he had forgotten where it was and referred the request to Rutherford. Rutherford said the data were scattered over several sites, not just one. Eventually Rutherford gave us a URL at Mann's FTP site, which pointed to a plain text file (pcproxy.txt), not an Excel file. After the controversy broke, we checked the date of creation of this file and found that it was the summer before, so it obviously wasn't a special collation generated for Steve. A few days after the publication of MM03, Mann apparently erased the file, removing the evidence of its age, but we had verified the date just before it was deleted. Now Mann was saying that the "right" data were at a different URL, which, despite several previous inquiries, he had never disclosed and we could not have accessed given the information we had to that date.

As for his suggestion that we had not noticed the errors, that was simply bizarre. Obviously we knew there were flaws in the data file we received—we spent twenty pages listing them! After finding the errors, Steve didn't keep using the flawed data; he rebuilt the whole data set from the original sources. Our reanalysis was based on the *corrected* data set, not the flawed data at the URL we had been given.

After we learned about the new URL, we compared it with the data from the old URL. Although the data at the old URL were collated, the data at the new URL were uncollated. We were quickly able to confirm that eighty-one proxy series, where there were no PC calculations, were identical in the two versions, and any comments in MM03 about these series were unaffected by the version dispute. We were also able to see what had happened in the collation at the old URL (pcproxy.txt). In the uncollated data, there were different versions of the PC series for each region, stored in subdirectories with names like BACKTO_1750, BACKTO_1600 and so on. Steve went through these folders series by series and was able to see how the collated data provided to us had been generated by having one column for each PC, with the values in the later subdirectory overwriting earlier values (subject to the collation error which we had noticed). On November 11 we posted our comments on the provenance of the data on our website.[6] Of course, that did not provide evidence one way or the other on what version was actually used in MBH98.

THE THREE KEY INDICATORS

A more substantive response by Mann and coauthors came a few weeks later in the form of a short paper by MB&H vigorously rejecting our conclusions.[7] They argued that our fifteenth-century results arose because we "selectively censored" early (pre-1500) segments of three "key indicators": a tree-ring width proxy from the site at Twisted Tree Heartrot Hill (TTHH) in northern Canada; the first principal component (PC1) of earlywood and latewood ring widths from a roster of ten sites in southwestern United States and Mexico (SW–M) studied by Stahle et al.; and the PC1 of ring widths and some densities from seventy-plus North American sites (NOAMER) partly overlapping the SWM network.[8] They presented a simulation showing that the early-fifteenth-century portion of their NH temperature index would, like ours, exceed the late twentieth century without these series in the fifteenth century. They argued that we had improperly deleted a substantial amount of early data, that our temperature index had a weaker statistical fit than theirs and accordingly, our results were worthless.

It was immediately obvious to us that the TTHH series was a red herring. The obsolete version used in MBH98 ended in 1976 and included a sixteenth-century portion based on a single tree; however, this version was not archived at WDCP. The archived site chronology goes up to 1992, but only commences in 1529, when three trees become available, and this is the version we used. However, even the TTHH series as used by MBH98 did not begin until 1459, so it is irrelevant to the pre-1450 interval in any case.

The issues with the two other indicators had arisen because of a problem in MBH98 disclosure. MBH98 said that they used 112 proxy series and "conventional" PC algorithms. Recall the question about how to handle missing data: Steve had decided to deal with it by calculating PCs over the maximum possible period. This led to different lengths for PC series than in pcproxy.txt, which we annotated in MM03 as one of the puzzles over the PC series in MBH98. Now Mann et al. said that they had used not conventional PC analysis, but a "stepwise" analysis in which they changed the rosters in each region in older periods, requiring 159 series instead of 112. The figure of 159 series had never been mentioned anywhere. It meant that there were PC series for the Stahle/SWM region and the NOAMER region in the AD1400 step of MBH98 calculations, which we did not have available in our calculations. This unavailability did not arise from "selective censoring," but from calculating PC series over the maximum period in which there were no missing data—which was our best interpretation of the obscure methodological descriptions in MBH98.

Replicating the stepwise method required a lot of information about which there was not a whisper in MBH98. First of all, what exactly were the 159 series? Mann et al. said that they had made fresh PC calculations for each region for each step, but that was clearly untrue, as this would lead to many more than 159 series. Also, at Mann's FTP site, there were directories for some periods, but not for others. Steve tried lots of different combinations, but it was impossible to come up with an exact match, balancing the right number of indicators in each period and still totaling 159 series. We asked Mann for details, but he refused to enlighten us, saying that we should be able to figure it out from his FTP site. Yet there were hundreds of PC series listed there, many of which were obviously not used. Eventually, Steve decided that the figure of 159 series was probably not correct (that still appears to be the case). But for the purpose of redoing our calculations back to 1400, Mann had stated that he used one PC from the Stahle/SWM network and two PCs from the NOAMER network, so we started with that. Steve made his best estimate of the numbers for the other regions by step. There turned out to be other errors in the MBH98 information that made these estimates a little off, but these discrepancies didn't matter for the main point in controversy, which now centered on the early fifteenth century.

In the end, none of this mattered for the SWM PC1. It turned out that its role could, as the lawyers say, be decided on alternate grounds. The SWM network is based on Stahle et al. (1998). For each site in the SWM network, MBH98 used two data series: earlywood and latewood widths, although Stahle et al. did not use latewood widths. Of the sites listed in the MBH98 Supplementary Information (SI) on the *Nature* website, only two of them (four series) are available before 1450. But Steve found that Mann's FTP site listed three sites (six series) extending back prior to 1450. This seemed to suggest a mysterious third site was used: but then two of the sites (four series) turned out to have identical values for the first 120 years for earlywood widths and the first 125 years for latewood widths, each differing thereafter. So they might have been spliced versions of different sites or different editions of the same site. Either way at least two series were clearly ineligible pre-1450, leaving only two potentially eligible sites. In other regions, MBH98 did not extend a PC1 back through an interval if only two sites were still available, and consistent application of this criterion would exclude the availability of the SWM PC1 in the pre-1450 period. Moreover, one of the two remaining sites is Spruce Canyon Colo., which is also in the NOAMER roster and should therefore have been dropped from the SWM group. The data for the remaining SWM site, Cerro Barajas, as used in MBH98, includes physically impossible negative values in the early portion of the series, which are not present in the version archived at WDCP. And Stahle et al. themselves did not apply their network prior to 1706. So on many grounds, we had reason to exclude the pre-1450 portion of this network. But there was an even more important reason. Even though Mann et al. had cited it as a "key indicator," our calculations showed that its presence or absence in the fifteenth century didn't matter. In our subsequent *Nature* correspondence, when we pointed out all these quality problems with this data, Mann et al. were only too happy to agree that it didn't matter for their early-fifteenth-century results.

The whole story thus seemed to turn on the North American PC1. We recalculated all our PC series following our estimates of stepwise inclusions, including the North American PC1 and PC2 (and the SWM PC1, which didn't matter) back to 1400. Using our recollated source data, we then recalculated the Northern Hemisphere temperature index. Our results still looked the same as MM03. So the difference was more than the recently revealed "stepwise" procedure for principal component calculations, as Mann was claiming, and it was linked somehow to the NOAMER PC1. Since the climate history of the world seemed to turn on this one indicator, we figured we'd find out everything we could about it. I'll return to this story in a moment.

In the meantime, Mann had also criticized other aspects of our emulation of his team's method. We had no interest in irrelevant disputes and figured we could settle the major methodological differences if we could just inspect the source code used in MBH98. Since Mann had redone his calculations in writing his response, we knew he had the source code at hand. We asked for it, but Mann refused to send it, or any more information for that matter. So Steve contacted the U.S. National Science Foundation (NSF), which had paid for the MBH98 research. Knowing there are rules requiring disclosure of NSF-funded data, we figured they might intervene. A program officer looked at our complaint and contacted Mann. Mann wrote back explaining what he had divulged to date and insisted he had fulfilled his NSF obligations. The file was eventually sent to the NSF general counsel, who ruled that there was no legal obligation for further disclosure: they regarded the code as Mann's personal property.

Unfortunately, we have since found this poor disclosure of data and methods is not an isolated situation in paleoclimatology. Other studies have an even worse record. Steve has contacted numerous paleoclimatologists in search of their data and has a thick file of excuses, dismissals, and brush-offs, along with a few honorable exceptions. Nor is the situation unique to paleoclimatology. Two economists recently took a 1999 edition of the *American Economic Review* (*AER*) and tried to replicate the empirical papers, only to find most authors unwilling or unable to share their data and command files in a usable format. So in 2004 the *AER* adopted a strict policy that empirical papers will no longer be published unless the authors supply their data and computational files to the journal's online archive.[9] More journals, even or especially paleoclimatological journals, should adopt a strict disclosure rule like *AER*'s. Advances in software and internet communication make this feasible and inexpensive.

We got caught up in another journal's debate on the subject in December, when we were alerted that a variation of Mann's response was listed on the *Climatic Change* website as "forthcoming." Steve wrote to the editor, Stephen Schneider, to protest about some of the language in it. Schneider wrote to assure us it was not, in fact, in press, but only under review, and indeed the version we saw was not the version actually being reviewed (there had been a mix-up at the *Climatic Change* office). He also said that he had mailed Steve an invitation to act as a referee. The new article referred to similar calculations as the ones that we had been trying to obtain information on. So in his new capacity as referee, Steve asked Schneider to obtain the supporting calculations and source code, so that he could carry out the requested peer review. That created consternation at *Climatic Change.* Schneider said that no one had ever made such a request in his twenty-eight years of editing of *Climatic Change* and refused to ask for that information without consulting his editorial board. The request prompted a long debate within the board (we were later told) about whether *Climatic Change* should ask its authors to release their computational files. The consensus was that it would be an excessive burden on reviewers if they were expected to review source code and supporting calculations. Steve replied that he was not suggesting that *Climatic Change* change its policies; he merely said that he was willing to examine the code in this particular case, and he did not see how that would set a precedent. Further, he could see no reason why *Climatic Change* should not request the information; after all, Mann might simply agree to provide it. Schneider still refused to ask for the source code, but agreed in February to ask Mann for the supporting calculations, that is, the results for the separate steps. In the end, Mann refused to provide the supporting calculations and Schneider asked that Steve complete his referee report anyway. Needless to say, the first comment in the referee report was that Mann et al. had failed to comply with the supposedly mandatory requirement to supply their data and supporting calculations. We haven't heard any more about this file.

THE FTP SITE

We knew that the whole MBH98 edifice now hung by the single thread of the NOAMER PC1 and that we had been unable to replicate the "key" PC1 using freshly collated data and the standard algorithm (princomp) in R. Steve now turned to Mann's recently unveiled FTP directory for MBH98, a strange, sprawling collection that had suddenly materialized in the wake of MM03. Prior to MM03, no reference to this URL had ever appeared, even in Mann's own web citations for MBH98 data. Mann's FTP directory contained data for no other paper.

The FTP site contained information on the series actually used in MBH98. Steve compared these listings with those in the original supplementary information (which he had used in his PC calculations) to see if this explained the difference in the NOAMER PC1. Discrepancies appeared immediately. Exactly 232 North American series were listed in the original SI, while only 212 series were actually used. He then checked the other networks and found similar problems: in the South American network, 18 series were listed in the original SI and only 11 were used in the actual calculations. There was no explanation why the series had been set aside, though an e-mail between the coauthors inadvertently left on the FTP site mentioned that deletion of the series arge030, one of the discrepant series, would be "better for our purposes."[10]

For the critical AD1400 step of the NOAMER network, the difference was only six series. We calculated the critical NOAMER PC1, using both the stated roster and the roster actually used. This did not make a material difference to the critical PC1, so the mystery of how to reconcile the NOAMER PC1 calculations remained.

While Mann continued to refuse to provide source code in response to our requests, in the folders at Professor Mann's FTP site we found remnant Fortran programs—perhaps unintentionally left on the site—that had been used to calculate the PCs. This was the only step in the entire calculation where we had the opportunity to actually inspect working source code. But we struck gold here.

Steve went through the Fortran programs line by line to see what they had done. Before doing a singular value decomposition (an algebraic factorization that yields PCs), they had "standardized" the series by subtracting the mean over the 1902–1980 subsegment, then dividing by the standard deviation over the same subsegment, then dividing again by the "detrended" standard deviation. In calculations using standard software, any standardization is done using a mean and standard deviation computed over the full length of the series, (say) 1400–1980 for the period in controversy, but in the MBH98 program they used the post-1901 portion of the data to compute the mean and standard deviation. This was not how the procedure had been described in MBH98. There they described subtracting the 1902–1980 mean and dividing by the 1902–1980 standard deviation prior to the regression module (i.e., later in the computation sequence). But they did not describe this procedure as having also taken place prior to the PC calculation. Since PCs are sensitive to changes in the way the data are standardized this would undoubtedly have raised the eyebrows of reviewers and readers, had they been told. It appeared that this had been done inadvertently, since an experienced user of PC methods such as Mann would have known that this would have an impact on PC calculations and that he would therefore have an obligation to disclose it. This apparently inadvertent error was at the heart of MBH98 and this, together with a little "trick" explained below, yielded Mann's famous hockey stick.

Steve also looked through the FTP site to see if there was any information on the results of the individual calculations for each step (especially the AD1400 step) which were used to support MBH's claims of high statistical accuracy, but he could not find any.

We remained frustrated with the Mann's obduracy in refusing to identify the 159 series or to disclose the results of his steps. Accordingly, in November 2003, we sent a "materials complaint" to *Nature,* under their policy requiring authors to disclose their data and methods.

The complaint included an item expressly noting the inaccuracy in MBH98 regarding the data transformation prior to the PC calculation, which we regarded as the issue most affecting the early fifteenth-century results. But we also listed other items that affected the integrity of the publication record, including the discrepancies between the series listed as being used and the series actually used. Sometimes, series were used twice in MBH98—with no notice to the reader or clear explanation. In the email mentioned above, one of the authors listed series that at a "wild guess" seemed to duplicate one another and recommended their removal, but, when we checked, his recommendations had not been carried out. While the NOAMER PC1 calculation accounted for most of the difference between MM03 (and now MM04) and MBH98, one of these duplications proved to account for the rest.

Another complaint was that, in a directory containing long temperature records, two series (nos. 6 and 8) were excluded without explanation.[11] Series 6 shows a conspicuously declining twentieth-century trend (figure 2.2). We hated to be suspicious, but the discrepancies were piling up. In total, we listed ten items where the disclosure in MBH98 appeared to be inaccurate; *Nature* forwarded this listing of ten items to MB&H for a response.

HOW TO MAKE A HOCKEY STICK

In January 2004, in addition to our materials complaint, we submitted an article to *Nature* that showed how the hockey stick was manufactured. We showed how the undisclosed programming error that Steve had discovered on the FTP site—subtracting the 1902–1980 mean (instead of the mean of the period of the principal component calculation (e.g., 1400–1980)—worked to pick out hockey stick–shaped series (if they were available in the network) and load them into the PC1.

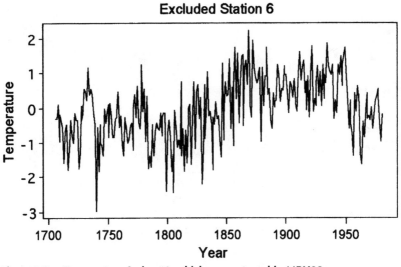

Excluded Station 6

Figure 2.2.   Temperature Series #6, which was not used in MBH98.

The error would not make much difference to networks that did not contain series with hockey-stick shapes (e.g., the Stahle SWM group). But the North American network did contain hockey-stick shaped series and the error had a big effect on the NOAMER PC1. For hockey-stick shaped series, the 1902–1980 mean is higher than the (say) 1400–1980 mean, sometimes significantly so. Subtraction of the 1902–1980 mean therefore inflated the variance in hockey stick–shaped series relative to what would happen from subtracting the 1400–1980 mean. Since PC algorithms load extra weight in the PC1 on series that have higher variance, the error resulted in picking out hockey stick–shaped series and overweighting them in the PC1. This was, presumably, a simple programming error, but it had a huge effect.

We showed this in a couple of different ways. First we showed the extreme differences in series weightings in MBH98 calculations. In the NOAMER roster for the AD1400 step, the most heavily weighted site is Sheep Mountain, Calif. (ca534). Sheep Mountain has a hockey-stick shape and Mann's algorithm gives it a whopping 390 times the weight in the PC1 of the least weighted series, Mayberry Slough, Ark. (ar052). The different shapes are shown in figure 2.3.

We'll call the Fortran program Mann1. Once we saw how the program worked, we tried an experiment to see if it could generate a hockey stick from random numbers. To generate the data we took the seventy

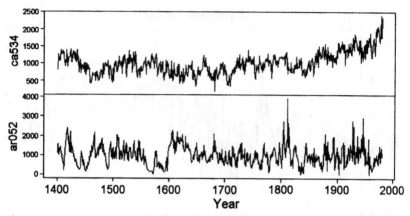

Figure 2.3.   Two NOAMER site chronologies used in MBH98. Because of the hockey stick shape, the first series is given 390 times the weight of the second series.

NOAMER sites available back to 1400 and fitted a lag-1 autoregression model to each. The coefficients ($\beta^1, \ldots, \beta^{70}$) were all of magnitude less than one. Then we generated seventy random vectors $a_t^1, \ldots, a_t^{70}$ of length 1081, using the AR1 formula $a_t^i = \beta^i a_{t-1}^i + e_t^i$, $i = 1, \ldots, 70$. Each series was initialized at zero and run using standard normal (N(0,1)) errors e_t^i. The first five hundred values were then dropped from each series, yielding seventy vectors of stationary red noise, each of length 581.

The (conventional) first principal component from these seventy series, after smoothing, showed the expected stationary sawtooth pattern (figure 2.4, top). Mann1 yielded the hockey stick–shaped PC1 shown in figure 2.4, bottom panel. The reason for the hockey-stick shape is that some of the underlying series randomly trail up or down at the end of their length, and these are selected for high weighting by the MBH98 method. Repeated experiments consistently returned hockey-stick shapes in the PC1.

Mann1 had a big effect on the AD1400 step of the North American network and we were able to show this led to the MBH98 version of the NOAMER network PC1, whereas a correct calculation led to our results.

One other error enabled a complete reconciliation of the differences between MM03-MM04 results and MBH98. Many series in MBH98 are duplicated within the data base. One of these, the Gaspé "northern treeline" series is used as a separate proxy (treeline 11) and in the NOAMER PC collation as cana036.[12] The data begin in 1404. When used as treeline 11, MBH98 gave the start date as 1400 and filled the empty first four cells by extrapolation (see MM03 table 5). This was the only extrapolation of a start date in the entire MBH98 corpus and we were consequently curious

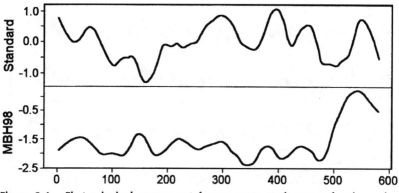

Figure 2.4. First principal component from seventy stationary red noise series, smoothed with Lowess (f=0.1). Top: Standard method (princomp command in the programming package R). Bottom: MBH98 algorithm, which first standardizes against the mean and standard deviation of the last seventy-eight values.

about it, especially now that the early-fifteenth-century results were at the heart of the controversy. The misrepresented start date enabled them to avoid disclosure of the unique extrapolation; the extrapolation enabled them to include this series in the AD1400 step, rather than withholding it until the AD1450 step. We calculated the NH temperature index without the extrapolation in the duplicate treeline 11 version and found that this seemingly innocuous four-year fill had a major effect on early fifteenth-century results (up to 0.4 degrees C in some years) and fully reconciled those differences between MBH98 and MM03–MM04, which were left after correcting the PC calculations.

We found other problems with the Gaspé series. Steve found that the first fifty years of the chronology fail standard minimum signal criteria.[13] The underlying dataset commences in 1404, but is based on only one tree up to 1421 and only two trees up to 1447. Dendrochronologists do not use site data where only one or two (or zero!) trees are available for generating a chronology. In fact the source authors don't use the series before AD1600 (see Jacoby and D'Arrigo 1989; D'Arrigo and Jacoby 1992).[14]

We now could carry out a comparison similar to the one in MM03, but based on a much closer replication of the MBH98 data and methodology, this time using the NOAMER PC1 and PC2 back to AD1400 (thereby answering the previous criticism from Mann et al.), but using correctly calculated PCs. The results are in figure 2.5, which was the conclusion to our submission to *Nature*. We submitted the article in January 2004, along with a cover letter explaining that it dealt with issues separate from

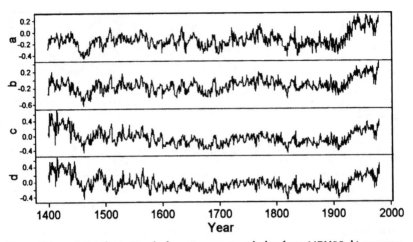

Figure 2.5.   a) Northern Hemisphere temperature index from MBH98; b) our emulation using publicly disclosed data and methods (correlation=0.89); c) reconstruction using correct PC calculation and excluding duplicate Gaspé series; d) applies other data updates and corrections. Methodology for bottom three panels is identical. Note particularly the change in relative height of the fifteenth-century portion of the graph.

our Materials Complaint, where we dealt with defective disclosure in MBH98, while the article dealt with impact of methodological errors (the methodologies also happening to be also undisclosed).

Figure 2.5's top panel shows the MBH98 graph, and the second panel shows Steve's best emulation, which has a correlation of 0.89. Since the replication also shows a unique late twentieth century and the basic hockey-stick form we were not concerned about the small remaining differences in result, and without full methodological disclosure by Mann there was nothing we could do about it anyway.

The third panel (c) shows the result using the same method as for (b), but applying correct PCs and removing the extrapolation of the duplicate Gaspé version. Those changes alone suffice to refute the conclusions in MBH98. The fourth panel (d) adds in all the data corrections as outlined in MM03 and the corrigendum of Mann et al.[15] Obviously the climate of the later twentieth century is unexceptional compared to the fifteenth century; to the extent this index summarizes the state of the climate.

At the end of February 2004 we received notice from *Nature* that, after consideration of the response by Mann et al. to our complaint, MB&H would be instructed to publish a corrigendum to MBH98 and be required to provide a new website with a listing of data and methodology sufficiently clear to permit replication of their results. We were specifically assured that the corrigendum would not engage in the controversy over the materiality of the errors but would simply list and correct them.

MB&H had also responded to our submitted article, and the papers were sent to two referees. Neither was convinced by the MB&H response and both supported publication. In March, we were asked to submit a revised version, responding to comments by referees and Mann et al. We were especially intrigued by an MB&H response point that our presentation exaggerated the effect of Sheep Mountain, since (they said) fourteen other series contributed heavily to the PC1. We wanted to know what the other fourteen series were, and this comment inadvertently proved to be a Rosetta stone for the final decoding of MBH98.

Almost all the NOAMER series selected for overweighting were of a single type and from a single researcher, Donald Graybill. The series were high-altitude bristlecone pine tree-ring chronologies, many of which had been studied by Graybill and Sherwood Idso as possible examples of CO_2 fertilization of tree growth, following a similar study by Lamarche et al. on Sheep Mountain.[16] The sites were selected for "cambial dieback," that is, the bark had died around most of the circumference of the tree. Graybill and Idso reported anomalously high twentieth-century growth for trees with cambial dieback, as compared with "full bark" trees at the same site. They reported that the anomalous twentieth-century growth was unrelated to the temperature data from nearby weather stations. In the case of Sheep Mountain, a weather station had operated within 10 km for nearly thirty years. Mann, Bradley, and Hughes themselves wrote in MBH99: "A number of the highest elevation chronologies in the western United States do appear, however, to have exhibited long-term growth increases that are more dramatic than can be explained by instrumental temperature trends in these regions." Later, coauthor Hughes in Hughes and Funkhouser (2003) would state that these elevated growth rates were a "mystery."[17]

Yet these sites were selected by Mann1 for such high weighting as to nullify the contributions of all other series in the NOAMER collection put together. None of the other proxies in MBH98 look like hockey sticks, but their influence was wiped out. The bristlecone and related series accounted for more than 99 percent of the weighting in PC1, which in turn was said to account for 37 percent of the variance in the North America network. In the subsequent regression calculation, the North America PC1 imparted its shape to the whole Northern Hemisphere temperature index, giving it the distinctive hockey-stick shape. This was startling enough, but there was an even bigger surprise to come.

At Mann's FTP site, there is a folder in the NOAMER directory called BACKTO_1400-CENSORED. You can imagine how intrigued we had been by this folder. It contained PCs, but it did not say what series were in it or what its purpose was. In another directory, Mann had a listing of 212 uncensored series and 192 "censored" series. Now that our attention had been drawn to the Graybill series, we checked to see if there was any connection to the twenty excluded sites. Again Steve struck gold. All fourteen of the Graybill sites were among the excluded series. The other six sites were also Graybill sites. We found that all twenty series were included among the seventy series in the uncensored BACKTO_1400 file, while there were only fifty series in the CENSORED file.

So now we knew what the calculation in the CENSORED file was— it redid the North American PC calculations in the critical AD1400 step after excluding the controversial Graybill sites, about which even Mann et al. were evidently worried.

This provided a test of our argument that the faulty PC algorithm was driving the MBH98 results. By removing these outlier series, the PC algorithm no longer had hockey sticks to overload on, and therefore should revert to a conventional shape. The comparisons appear in figure 2.6. The top

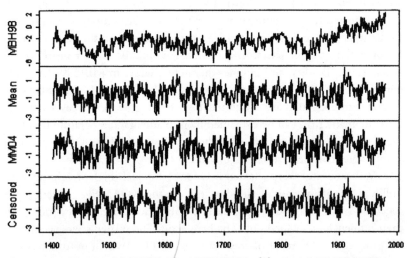

Figure 2.6.   Top panel (MBH98): The MBH98 PC1 of the post-1400 NOAMER tree-ring network, computed using 1902–1980 subsegment standardization. **Second panel (Mean):** simple mean of proxies. **Third panel (MM04):** PC1 as computed by standard software standardizing over the full period. **Bottom panel (Censored):** Unreported PC1 computed by MB&H after censoring Graybill-Idso high-altitude series.

panel shows the NOAMER PC1 as computed by MBH98 using their erroneous standardization. The resemblance to the Sheep Mountain site (figure 2.3a) is evident. The second panel shows the simple mean of the seventy proxies in question. This is quite similar to the third panel, which shows the first PC of the same seventy series correctly computed, that is, after standardizing over the full period. The bottom panel in figure 2.6 shows the PC1 from the CENSORED file, that is, as computed by MB&H themselves after dropping the twenty Graybill sites. It matches the correctly computed PC1 (MM04) almost perfectly. Evidently we weren't the first ones to discover the role these bristlecone series were playing! Mann had also wondered about them, and redid his calculations without them. Given his subsequent furious excoriation and accusations of "selective censoring," we were amused to find results similar to ours lurking on his FTP site in a folder called "CENSORED."

I must note the hilarious irony in the fact that data published (in part) by Sherwood Idso, the famous global warming skeptic, would, as the result of someone's programming error, reappear many years later as the NOAMER PC1, now billed as the "key indicator" and principal evidence for the IPCC's position on global warming! We submitted our response incorporating these comments on March 18.

Meanwhile, on March 16, we received a page proof of the corrigendum. It left out the most important inaccuracy of MBH98—the incorrect disclosure of the (incorrect) PC methodology, as well as many other items. It even contained a few new errors of its own.

The corrigendum attributed the discrepancy in the series listings to the application of quality control rules additional to those already mentioned in Mann et al. (2000),[18] which included the following criteria: a tree-ring chronology would not be used unless the mean correlation of individual tree-ring records with the site chronology was at least 0.5; the chronology commenced by 1626; and it was composed of at least eight tree-ring segments by 1680. Steve had previously gone through the MBH98 data and found thirty-nine series that did not commence by 1626, twenty-two sites that did not have eight trees by 1680, and 171 sites that had less than 0.5 mean correlation of the individual trees with the site chronology. In one of the cases (cana153) the tree-ring segments had so little correlation with the site chronology that Steve emailed the originating author (Roseanne D'Arrigo) to ask why. She discovered that the wrong data had been posted at WDCP and immediately asked them to remove the listing. That surely would have been noticed earlier if Mann had indeed applied his quality control rules to all the data.

We wrote to *Nature* to object all of these matters and they temporarily pulled the corrigendum out of production. About ten days later, however, *Nature* dismissed our concerns about the inaccuracies in the purported explanation of the discrepancies, on the grounds that they were "irrelevant" to our original materials complaint. *Nature* also said that "space limitations" precluded a listing of all the errors, but they felt confident that the new SI would provide a complete record of the data actually used. On the same day that our objections to the draft corrigendum were dismissed because of "space limitations," we were asked to cut our article down to eight hundred words to fit the format of a Communication Arising, which we did on April 9, 2004.

The corrigendum was eventually published on July 1, 2004, as was a new online data archive for MBH98. The printed version contained the preposterous claim that none of the errors had any effect on the results, a claim that had been inserted after production of the page proofs we were shown, and which was obviously inconsistent with material then under review at *Nature*.

Almost four months later (August 4) we were told our article would not be published. *Nature* said the principal reason was that the material could not be adequately explained in the (now) five hundred words they said would be available for our paper.

We did not have an opportunity to respond to the second round of comments from Mann and his coauthors. One counterargument seemed to impress one of the referees. Mann et al. acknowledged the data transformation prior to computing the PCs, but claimed that, even using corrected PC calculations, they could get MBH98-like results if they expanded the list of retained PCs in the AD1400 step for the NOAMER roster to five, instead of only two as in MBH98. They pointed out that the bristlecone pines get heavily weighted in the PC4. For their regression module, it doesn't matter whether the series appear as a PC1 or PC4; either way the bristlecone pines still lever the final NH temperature index into a hockey stick. Without the NOAMER PC4 (which explains less than 8 percent of the variance of the NOAMER network), they would end up with the same results as we got; by including it, they get results that resemble MBH98. So their key conclusion hinges on ginning up a reason to include a minor PC representing the growth of trees that numerous experts have said are not good proxies for temperature.

Ultimately the issue is robustness. Mann et al. made grandiose claims about the "robustness" of their methods—even claiming in Mann et al. (2000) that their method was robust to the exclusion of all tree-ring information. Yet it is not even robust to the presence or absence of the bristle-cone pines or to the NOAMER PC4.

In addition to loading hockey-stick shapes into the PC1, the programming error also made the bristlecone "signal" in the PC1 appear to be far more dominant than it really was. It attributed 37 percent of the explained variance in the entire network to the (incorrect) PC1, thus suggesting the Graybill-Idso sites were the dominant pattern for the whole continent, while the correctly calculated PC4, to which these sites actually get weighted, accounts for less than 8 percent of the variance in the NOAMER network. Thus, what was erroneously argued to be the "dominant" signal in the entire Northern Hemisphere climate turns out to have been a local phenomenon specific to a group of high-altitude bristlecone pines, whose influence was inflated due to a programming error. Mann's hockey stick hinges (literally) on this. And on that flimsy foundation the Intergovernmental Panel on Climate Change based the conclusions of its third assessment report.

The MBH98 data, if analyzed using MBH98 methods, but without erroneous principal component calculations and without the duplicate Gaspé data versions (or even without the undisclosed extrapolation in the duplicate Gaspé series), does not support the conclusion that the twentieth-century climate is unusually warm and does not enable any far-reaching conclusions about where the 1990s rank in millennial temperature.

CONCLUSION: PEER REVIEW
AND THE POLICYMAKING PROCESS

Canada's Chief Climate Science advisor, Henry Hengeveld, recently dismissed our work in the *Canadian Meteorological and Oceanographic Society Bulletin* by arguing that we are not experts and therefore not competent to identify errors in MBH98. If so, just think what some real "experts" could have discovered had they seriously scrutinized MBH98. But of course the real "experts" did not bother. Despite its spellbinding role in international science and public policy, other than our efforts there was no due diligence on Mann's hockey stick.

Steve McIntyre likes to contrast what we've witnessed with the layers of due diligence involved in even small offerings of securities to the public. A prospectus must contain audited financial statements. Auditing is carried out by specialized and highly paid professionals and, for large corporations, the audit is virtually a full time occupation. A company issuing an exploration prospectus must provide a qualifying report on its geological properties by an independent geological professional. The geologist must be truly independent. A person signing a prospectus could not use his own reports as "independent" reports. Both the auditor and the independent geologist must approve the relevant language in the prospectus and provide signed consent letters to the securities commission. The prospectus itself is reviewed by two sets of securities lawyers—one for the issuing corporation and one for the underwriter or broker acting as agent. Then the prospectus is reviewed by the securities commission, a nit-picking process. Any errors identified by or concerns of the securities commission must be dealt with, regardless of whether it "affects the results." The process is expensive and painstaking. After all this, the officers and directors have to sign a form certifying that they have made "full, true, and plain" disclosure, which means not only certifying that everything in the prospectus is true to the best of their knowledge, but also that they have not omitted anything from the prospectus that is material. Even small public offerings have multiple layers of due diligence.

Despite the multiple layers of due diligence for prospectuses, frauds still occur. Lots of people believed in Bre-X and Enron, including people as eminent in their fields as those currently supporting IPCC, and they turned out to be wrong. In both of those cases, there were lapses in due diligence. In the case of Bre-X, the drill core was famously never available for inspection. During its main boom, Bre-X never issued a prospectus. When it listed on the Toronto Stock Exchange, it filed an ore reserves study by a well-respected and eminent engineering firm, which contained the caveat that the ore reserve calculation relied on company information and that no examination of drill core or verification were carried out. The fraud was immediately exposed when the first third-party drill core was done.

In the case of Enron, in retrospect, it seems that analysts never really knew what Enron did or how it made its supposed profits. In both cases, and in common with the dot.com boom, there was a "madness of crowds." Those happen from time to time in public markets despite the best efforts of analysts and regulators. While these examples may seem very foreign to the academic world, how does due diligence for MBH98 bear up in comparison?

First, "peer review" for an academic journal is a much lesser form of due diligence than an audit of financial statements. The referees of our submission expressly stated that attempting to determine who was right or wrong as between MBH98 and ourselves was far beyond the scope of review that they could provide. Yet the differences are important to public policy and the resolution of these issues should be quite routine, if the original authors were required to cooperate. Auditors would definitely resolve this matter in a business situation. In our dealings with *Nature,* even where issues were explicitly identified, peer reviewers were unable to provide sufficient due diligence to resolve the matter. We are quite confident that *Nature's* peer reviewers for the original publication did not examine the data or the programs used to produce Mann's hockey stick or carry out any audit level due diligence.

At the IPCC level, the IPCC itself made no attempt to verify any MBH98 findings, relying only on the prior peer review by *Nature.* There is a common misunderstanding by the general public and the numerous Nobel laureates who endorsed the IPCC report that the IPCC carried out substantial due diligence of its own. That is not the case. Obviously, problems can result if people think that due diligence has taken place when it hasn't.

The failure of the IPCC to carry out such independent verification or to audit studies may be partly explained by the lack of independence between the chapter authors and the original authors. Michael Mann was lead author of the chapter relying on his own findings, a lack of independence that would never be tolerated in ordinary public offerings of securities.

Prior to our work, there was no effort by any paleoclimatologist to specifically replicate MBH98. One paper copied the method to explore how simulated proxy data might compare to simulated temperature data in a climate model simulation, but did not attempt to reproduce the MBH98 results.[19] There are other multiproxy studies arriving at somewhat similar conclusions as MBH98, but most of these studies are not truly "independent." The most often cited multiproxy studies are nearly all by a small group of coauthors: Mann, Bradley, and Hughes (1998, 1999), Mann and Jones (2003), Jones and Mann (2004), Bradley and Jones (1993), Jones, Briffa, et al. (1998), Briffa, Osborn, et al. (2001), and so on, and many reuse the same basic underlying data.[20]

In most branches of science, specific replication is required before results are accepted. Yet in paleoclimate, the idea of our merely trying to replicate MBH98's findings has been derided by many climate scientists. Apart from expressing scandal that we didn't just take the findings on authority, they argued that we should have developed our own proxies and produced our own index. But whenever someone proposes a new method, involving advanced methods and a great deal of data handling, if the results are deemed canonically significant it seems self-evident that the programs should be checked to see that there were no errors or unstated methodological variations. No one bothered.

One simple suggestion to minimize similar problems and facilitate due diligence is this: to make independent replication possible, the data and the methods must be published in unambiguous form. Rules such as those now in force at the *American Economic Review* should be universally applied in paleoclimatological publications in the future. In dealing with the backlog of poorly documented papers, institutions seeking to apply older papers should determine whether they have met full disclosure standards prior to citing the papers, and any shortcomings in the availability of complete details on data and methodology should be prima facie grounds to forbid a paper's use in public sector decision making.

Beyond that, there is an obvious need for additional due diligence prior to use of academic articles in public policy. In the private sector, no one would build an oil refinery based on an academic article. There is a process of engineering due diligence. Some of the most highly paid professionals are principally involved in verification. Yet governments will make far larger, costlier decisions based on the chimerical standard of academic peer review. Merely stating the contrast points to the need to ramp up standards in the public sector, and quickly.

NOTES

1. M. E. Mann, R. S. Bradley, and M. K. Hughes, "Global-scale Temperature Patterns and Climate Forcings over the Past Six Centuries," *Nature* 392 (1998): 779–87.

2. M. E. Mann, R. S. Bradley, and M. K. Hughes, "Northern Hemisphere Temperatures during the Past Millennium: Inferences, Uncertainties and Limitations," *Geophys Res Lett* 26 (1999): 759–62.

3. M. E. Mann and P. D. Jones, "Global Surface Temperature over the Past Two Millennia," *Geophys Res Lett* 30 (2003): doi: 10.1029/2003GL017814; P. D. Jones and M. E. Mann, "Climate over the Past Millennium," *Rev Geophys* 42 (2004): RG2002, doi:10.1029/2003RG000143.

4. S. McIntyre and R. McKitrick, "Corrections to the Mann et al. (1998) Data Base and Northern Hemispheric Average Temperature Series," *Environment and Energy* 14 (2003): 751–71.

5. ftp://holocene.evsc.virginia.edu/pub/MBH98/ (hereafter *FTP*).

6. http://www.uoguelph.ca/~rmckitri/research/trc.html.

7. ftp://holocene.evsc.virginia.edu/pub/mann/EandEPaperProblem.pdf.

8. D. W. Stahle et al., *Bull Amer Meteorol Soc* 79 (1998): 2137–52; D. W. Stahle and M. K. Cleaveland, *J Climate* 6 (1993): 129–39.

9. See the editorial statement in *AER,* March 2004, 404.

10. ftp://holocene.evsc.virginia.edu/pub/MBH98/TREE/VAGANOV/ORIG/malcolm_29-JUL-97.

11. ftp://holocene.evsc.virginia.edu/pub/MBH98/INSTR/TEMP.

12. This series was included in the North American northern treeline network. The Gaspé peninsula is nowhere near the northern treeline.

13. T. M. L.Wigley, K. R. Briffa, and P. D. Jones, "On the Average Value of Correlated Time Series with Applications in Dendroclimatology and Hydrometeorology," *J Clim App Meteor* 23 (1984): 201–13.

14. G. C. Jacoby and R. D. D'Arrigo, "Reconstructed Northern Hemisphere Annual Temperature since 1671 Based on High-latitude Tree-ring Data from North America," *Clim Change* 14 (1989): 39–59; R. D. D'Arrigo and G. C. Jacoby, "Dendroclimatic Evidence from Northern North America," in *Climate since A.D. 1500,* ed. R. S. Bradley and P. D. Jones (New York: Routledge, 1992), 246–68.

15. M. E. Mann, R. S. Bradley, and M. K. Hughes, "Corrigendum," *Nature* (2004): doi:10.1038/nature02478.

16. D. A. Graybill and S. B. Idso, "Detecting the Aerial Fertilization Effect of Atmospheric CO_2 Enrichment in Tree-ring Chronologies," *Global Biogeochem Cycles* 7 (1993): 81–95.

17. M. K. Hughes and G. Funkhouser, "Frequency-dependent Climate Signal in Upper and Lower Forest Border Trees in the Mountains of the Great Basin," *Climatic Change* 59 (2003): 233–44.

18. Michael E. Mann, Ed Gille, Raymond S. Bradley, Malcolm K. Hughes, Jonathan Overpeck, Frank T. Keimig, and Wendy Gross, "Global Temperature Patterns in Past Centuries: An Interactive Presentation," *Earth Interactions* 4 (2000); http://www.ngdc.noaa.gov/paleo/ei/ei_nodendro.html.

19. E. Zorita, F. Gonzalez-Rouco, and S. Legutke, "Testing the Mann et al. (1998) Approach to Paleoclimate Reconstructions in the Context of a 1,000-year Control Simulation with the ECHO-G Coupled Climate Model," *J Climate* 16 (2003): 1378–90.

20. R. S. Bradley and P. D. Jones, "'Little Ice Age' Summer Temperature Variations: Their Nature and Relevance to Recent Global Warming Trends," *The Holocene* 3 (1993): 367–76; P. D. Jones, K. R. Briffa, T. P. Barnett, and S. F. B. Tett, "High-Resolution Paleoclimatic Records for the Last Millennium: Interpretation, Integration and Comparison with Circulation Model Control-run Temperatures," *The Holocene* 8 (1998): 455–71; K. R. Briffa, T. J. Osborn, F. H. Schweingruber, I. C. Harris, P. D. Jones, S. G. Shiyatov, and E. A. Vaganov, "Low-frequency Temperature Variations from a Northern Tree Ring Density Network" *Journal of Geophysical Research* 106 D3 (2001): 2929–41.

3

OBSERVATIONAL SURFACE TEMPERATURE RECORDS VERSUS MODEL PREDICTIONS

Robert C. Balling Jr.

The overwhelming majority of global warming concern comes from the output of theory-based numerical climate simulation models developed and run throughout the world. In virtually every case, from the simplest models developed as early as the nineteenth century (Arrhenius 1896) to the most sophisticated ones in use today, the robust and highly publicized prediction is that the continued buildup of greenhouse gases will cause accelerated global warming (Houghton et al. 2001). Even when many other factors are considered—the effects of sulfates, solar variability, ozone depletion, dust, land-use changes, cyclical El Niño events, periodic volcanic eruptions, jet contrails, and all the rest—the numerical models continue to show a world that warms rapidly. Those simulations, simply using changes in the greenhouse effect, generally suggest that the planetary near-surface air temperature should have increased by approximately 1.5°C over the past one hundred years and that the temperature rise will continue at a higher rate for the next century.

Climate scientists from around the world have gathered surface and near-surface temperature measurements, using them to verify model projections that global temperatures are on the rise, that the rate of that temperature rise is increasing, that most of the record-breaking global temperatures have occurred in the recent period, and that these unusually high global temperatures are internally consistent with greenhouse warming. More than a few leaders around the world have declared that global warming is no longer a theoretical prediction, but a fact confirmed by observations throughout the planet. The debate becomes infinitely more complex given the relative lack of warming in recent decades in the lower tropospheric balloon-based and satellite-based temperature records (Wallace et al. 2000). To keep our discussion simple, we will limit ourselves to considering the relationship between observed and predicted temperatures measured at or near the surface of the earth. John Christy of the University of Alabama discusses these complicated satellite temperatures at length in the next chapter.

DEFINING PLANETARY TEMPERATURE

At first glance, it would seem easy to determine whether the planetary temperature has been increasing over the past hundred or so years since many weather records extend for more than a century. In theory, we should be able to assemble those records, check for trends, and determine whether or not the world is generally warming, cooling, or remaining near the same temperature. That exercise has indeed been carried out, and the world is now warming at a rate that is "not inconsistent" with the expectations from current model simulations, to borrow the artful phrase first coined in 1985 by the Department of Energy in an analysis of climate science.

Although other global temperature records are available from the thermometer network of the world (e.g., Hansen and Lebedeff 1987; Vinnikov et al. 1990), the most popular and widely used record has been developed and maintained by Phil Jones of the Climate Research Unit at the University of East Anglia (Jones 1994, and updates). His data set is based on the records of several thousand land-based stations and millions of weather observations taken at sea (Parker et al. 1995). Jones has carefully assessed the homogeneity and representativeness of each time series, and made an effort to identify and eliminate errant values. He converts the monthly station observations into 5° latitude by 5° longitude grid-box data, and all values are expressed as deviations (anomalies) from a reference period defined as 1961–1990. The grid box anomalies may then be areally averaged for each hemisphere, and the two hemispheric values are in turn averaged to determine the estimate of global temperature.

A plot of the annualized values shows considerable support for the global warming prediction (figure 3.1), although the observed magnitude is approximately one-half of what is obtained from climate models that simply change the greenhouse effect (which is why a concurrent cooling agent—sulfate aerosol—began to be parameterized in models about a decade ago). Over the entire 1900–2002 period, the near-surface air temperatures rose linearly by 0.069°C decade^{-1}; warming spurts occurred from

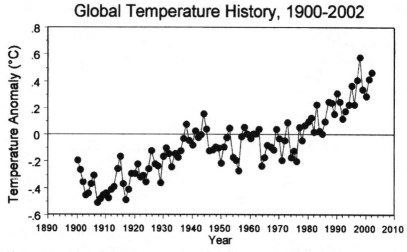

Figure 3.1. Plot of annual near-surface global temperature anomalies over the period 1900–2002.

the late 1910s to 1945 and from 1970 to the present. Since 1970, the warming rate of the earth has accelerated to 0.17°C per decade, about three times as fast than the average rate for the past century. However, this comparison (while often made) is somewhat misleading because there was a slight cooling from 1940 to 1970. The actual rates of warming in both eras (1915–1945 and 1970–2100) are very close, at 0.16°C and 0.17°C per decade, respectively.

The fifteen warmest years on record all occurred from 1983 onward, and the twenty coldest years all occurred before 1930. In the eyes of many, this is the smoking gun in the greenhouse debate: Statistically significant warming appears in the global near-surface air temperature records, and that warming has accelerated in recent decades. But, in fact, it is not any different than the rate of warming that occurred much earlier in the century, before there were any appreciable changes in the greenhouse effect.

The geographic pattern of the recent warming is broadly consistent with model simulations as well. One consistent prediction from the models used by the Intergovernmental Panel on Climate Change (IPCC) is that the cold and dry air of the high-latitude northern hemispheric locations will be particularly responsive to the increase in greenhouse gases (figure 3.2). In lower latitudes, the higher levels of atmospheric moisture produce such a strong natural greenhouse effect that additional concentrations of carbon dioxide have less of an effect compared with the drier air masses. As figure

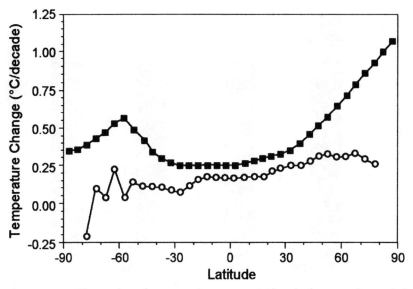

Figure 3.2.   Observed surface warming rate (°C decade⁻¹) over the period 1970–2002 (open circles) vs. predicted warming rate (filled squares).

3.2 shows, the high latitudes of the Northern Hemisphere have warmed more than any other location on the planet, even though the warming rate is far less than what has been predicted by numerical models.

The geographic patterns may be broadly consistent from numerical models to actual observations, but is there a seasonal fit as well? Apparently so. Models consistently predicted the greatest warming in the winter season (December–February in the Northern Hemisphere and June–August in the Southern Hemisphere) and the least warming in the summer season (figure 3.3). When the global temperature is calculated by season, there is indeed a slight tendency for the winter to have the greatest warming, but the observed seasonality to the warming in the 1970–2002 period is less than what the models predicted. Although the Jones temperature data are monthly averages and do not allow any analysis of the diurnal patterns of the warming rates, many other scientists (e.g., Easterling et al. 1997) have conducted analyses of daily maximum and minimum temperature data, and even hourly temperature data, and have identified a clear global signal of greatest warming at night and least warming during the day. This diurnal pattern in the warming rates is evident in many numerical models used to simulate the effect of increasing concentrations of greenhouse gases (Rind et al. 1989).

When added together, the evidence from the near-surface air temperature measurements seems totally convincing that the planet is responding

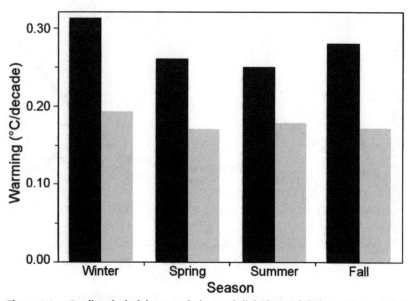

Figure 3.3.   Predicted (dark bars) and observed (light bars) global warming rate (°C decade⁻¹) over the period 1970–2002 by season.

to the buildup of greenhouse gases. While the global trend is too low for a model greenhouse-only response, both the regionality and seasonality of the warming are broadly consistent with the model simulations; and the diurnal pattern of the warming is broadly consistent with expectations given the models' output. We could stop here and declare, as so many others have, that the observations are consistent with the theory, and therefore the science is settled. But there is a lot more to this story.

HOW RELIABLE ARE THE RECORDS?

Missing Data

One of the problems with the near-surface air temperature record is that substantial parts of the globe lack the measurements needed to generate a monthly temperature anomaly for various 5° latitude by 5° longitude grid cells. Ocean areas off of major shipping lanes, ice-covered areas, many arid and hyperarid areas, and mountainous areas often lack the temperature records necessary to generate the monthly temperature anomalies for individual grid cells. And not surprisingly, the area of the earth without valid data increases further back in time and also during periods of global strife. Less than 30 percent of the planet had temperature records at various times in the twentieth century, and today, fully 20 percent of the earth is not covered by the Jones database (figure 3.4). While this poses substantial problems for generating an accurate trend over the past hundred-plus years, it is noteworthy that the global temperature has increased substantially during the past three decades, a period when coverage has hovered near 80 percent. Furthermore, Karl et al. (1994) produced some interesting calculations on how the pattern of missing data in various 5° latitude by 5° longitude grid cells impacts the global temperature anomalies and trends in the resultant time series. They concluded that the temporal pattern in missing values would likely produce only small impacts on the global temperature trend. This is confounded by the fact that the longest, most reliable records tend to originate in the land areas of the Northern Hemisphere, particularly North America, Europe and the former Soviet Union. These areas should show enhanced warming under a changing greenhouse, so the fact that the observed greenhouse-only warming is still beneath the average global projections for the models remains somewhat troubling to the glib consensus of model rectitude. Amount of biasing from missing data: unknown.

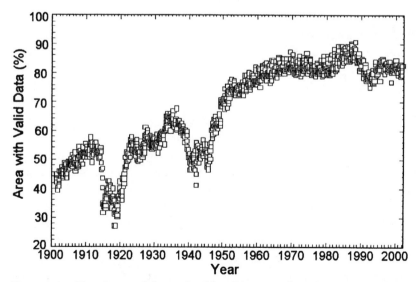

Figure 3.4. Plot of area of the earth with valid near-surface air temperature data from 1900–2002.

Urban Heat Island

One of the main potential problems with the land-based temperature record is that many meteorological stations are located in growing urban areas and the climatic effects of the urbanization process may overwhelm any signal related to the buildup of greenhouse gases. The physical causes for localized warming in urban areas have been studied vigorously for more than a century (e.g., Howard 1833), and these causes are reasonably well understood. Although the city size, morphology, relief, elevation, and regional climate will determine the intensity, persistence, and configuration of the heat island, several generalizations are possible regarding the causes of the localized warming.

The principal reason for the development of most urban heat islands is the waterproofing of the urban surface. In many cities, the natural vegetation is largely removed, and the surface covered by nearly impervious materials. Precipitation quickly runs off the urban surface into underground storm sewers, and the resulting surface moisture and near-surface moisture in the city is minimized. Given reduced quantities of moisture, large portions of the net radiation at the surface go toward heating the surface and air as opposed to evaporating or transpiring water. Other reasons for the warmer conditions in cities include the thermal properties of building materials; urban canyons that reduce the loss of long-wave terrestrial radiation by narrowing the sky-view factor; the actual release of heat from a variety of anthropogenic activities within the city; and the trapping of long-wave terrestrial radiation by low-level atmospheric aerosols.

There is no doubt that urbanization leads to warming in most cities, and there is little question whether this warming has contaminated the near-surface temperature records that are used to create the global temperature estimates. Many clever studies have been conducted using pairs of stations from urban and rural areas, statistics on population trends, intensity of night light as measured from satellites, and even high-resolution land-use maps in an attempt to quantify the urban heat island effect on the global temperature time series (Karl et al. 1988; Balling and Idso 1989; Peterson et al. 1999; Gallo et al. 1999). Although the exact impact differs with the methodology used, it appears that urbanization has added a warming rate near $0.006°C$ decade^{-1} (Houghton et al. 2001) representing approximately nine percent of the global warming rate of $0.069°C$ decade^{-1} over the 1900–2002 time period.

Desertification

In many dryland areas of the world, overgrazing and other human-caused activities have degraded the landscape, substantially reducing the amount of biomass in those areas. The reduced vegetative cover has a variety of effects, but the largest impact in terms of near-surface air temperature is the reduction of soil moisture. Once plants are removed, runoff accelerates, surface friction is reduced and the near-surface wind speeds increase, the surface temperature increases, and the rate of potential evapotranspiration increases. All of these factors reduce soil moisture and produce feedbacks that accelerate the desertification processes (Williams and Balling 1996). Just as in cities, the reduction in surface moisture means that less net radiation goes to evaporating and transpiring water and more radiation is used to heat the surface and heat the air. The end result is a warming signal in many dryland areas that could be inadvertently ascribed to the buildup of greenhouse gases. Although this effect can be significant at local and regional scales, it is likely that the desertification effect contributes only about 0.01°C to the global temperature rise (Nasrallah and Balling 1994).

Instrument Problems, Adjustments

The search for the global warming signal is further complicated by a number of other potential contaminants to the historical temperature records. The traditional mercury-in-glass thermometers are being replaced worldwide by electronic thermistors capable of continuous air-temperature monitoring. The change in instruments may be introducing a positive (i.e., warming) bias into the record. During the summer season, turbulent eddies of air pass by instrument shelters in any convectively active atmosphere. In the past, the warm eddies would have passed by the glass thermometers with insufficient time to heat the glass and the mercury inside. But the new, highly sensitive electronic sensors immediately recognize and record the temperature of each turbulent eddy, and the result inflates the maximum daily temperatures in the recent period (Gall et al. 1992). The shelter itself introduces yet another warming trend related to its fresh white paint, which deteriorates over time, thereby changing the shelter's reflectivity and ultimately the temperature within the shelter.

The problems hardly stop there. In many parts of the world, initial settlements occur in valleys, often near rivers, where cold air drainage can keep minimum daily temperature relatively low. As more stations are established away from the river valleys, rural airports are specifically chosen as sites because they minimize cold air drainage (which leads to preferential fog formation). In many cases, the regional temperature will appear to move spuriously upward as fewer stations are impacted by the effects of cold air drainage. Another problem deals with the time of observation bias: For much of the twentieth century, an observer would typically read the maximum and minimum thermometer in the morning, often near 0700 LST. Notice that during an unusually cold morning, the low minimum temperature would be double-counted. The observer records that low value for the previous twenty-four hours, resets the maximum and minimum thermometer, and immediately records the unusually low temperature as the minimum value for the next twenty-four hours. With the introduction of electronic sensors, on the other hand, monitoring is nearly continuous, and the double-counting of extreme low temperatures is eliminated, thereby introducing a warming bias in the record. Balling and Idso (2002) used balloon-based and satellite-based lower-tropospheric temperatures to show that the adjustment for the time of observation bias was introducing "a nearly monotonic, and highly statistically significant, increase of over 0.05°C per decade" in the widely used U.S. temperature database. Because the spread of electronic thermistors is primarily in the past three decades and mainly in the industrialized and developed world, a reasonable upper limit for its effect on overall global warming is probably around 0.10°C to 0.15°C.

Clearly, such inadvertent, but very real, warming biases can and do compromise the integrity of the widely used temperature records of local, regional, hemispheric, and global scales. In combination, they could artificially inflate the observed warming by 0.2°C to 0.3°C.

With 71 percent of the earth covered by ocean, any discussion of temperature bias must acknowledge the special problems of measurements over the water. Many marine "surface" temperatures are measurements of water, not the air adjacent to its surface, the assumption being that the surface water temperatures would be close to the near-surface air temperatures, particularly at night. Indeed, many decades ago, mariners would measure sea-surface temperature after pulling canvas or wooden buckets onto the deck and inserting a glass thermometer into the captured sea water. Later, mechanical injection systems for cooling water were used to measure sea-surface temperature, introducing an obvious warm bias from the ship itself. Some vessels had weather stations on board to measure actual air temperature; of course, a weather station on the deck of the ship would be impacted by its positioning with respect to other features on the deck and the actual height of the deck above the sea surface. Such limitations introduce substantial biases into the ocean records (Parker et al. 1995). More important, Christy et al. (2001) concluded that trends in sea-surface temperatures are not the same as trends in air temperature and that interchanging sea-surface and actual air temperature measurements may in fact compromise the depiction of temperatures through time.

The problems with the historical temperature records notwithstanding, it is very likely that the recent upward trend in figure 3.1 is very real and that the upward signal is greater than any noise introduced from uncertainties in the record. Still, the general error is most likely in the positive direction, with a maximum possible (though unlikely) value of 0.3°C. Even so, that all biases are positive implies that the indicated rise of 0.7°C is nonetheless an overestimate. And consider that globally averaged retreats of mountain glaciers, decrease in spring snow cover, a decrease in sea ice, and even an observed increase in water vapor are all generally consistent with a near-surface warming in recent decades (Houghton et al. 2001). It is certainly tempting to conclude that the warming in the historical near-surface air temperature record is a response to the buildup of greenhouse gases. But many other possible explanations exist.

ALTERNATIVE EXPLANATIONS
FOR TRENDS AND VARIATIONS

Little Ice Age

The warming of 0.16°C per decade that maximizes in the 1915–1945 period is difficult to ascribe to changes in the greenhouse effect. Its magnitude is essentially the same as the subsequent warming of the past three decades, yet the human greenhouse "forcing" was much smaller—from approximately 10 percent of the current change in 1915 to about 35 percent by 1945. Given that much of the measured warming of the past one hundred years is real, we are still left with the question of whether that warming is related to the buildup of greenhouse gases over the same time period. Rather obviously, the planetary temperature has risen and fallen many times in the past when humans had no chance to alter the climate at any temporal or spatial scale. Given the massive swings in global temperature approaching 10°C from coldest periods to warmest periods, the approximate 0.70°C temperature rise since 1900 is nothing out of the ordinary when viewed over long periods of the earth's history.

The last great ice age ended approximately 12,000 years ago, and we remain in an interglacial warm period that, based on orbital forcing of global climate, should last another four thousand to five thousand years. However, within warm interglacial periods, the hemispheric and global climate system periodically endure substantially cooler times. About 550 years ago, the European, northern hemispheric, and planetary temperature fell by a few degrees, leaving the earth in the Little Ice Age, a period ending in the mid-nineteenth century (Grove 1988). While the global nature of this event has been challenged (and the challenge rather uncritically accepted by the United Nations Intergovernmental Panel on Climate Change) by Mann et al. (1998, 1999), a recent comprehensive study of paleoclimatic indicators by Soon and Baliunas (2003) appears to have permanently proven the global realism of the phenomenon. The warming evident in the thermometer-based near-surface air temperatures in the first part of the twentieth century may be little more than a natural recovery from the Little Ice Age.

The Little Ice Age was initially recognized in northern Europe when scientists began studying alpine glacial remains and reviewing historical records for the fifteenth-to-nineteenth centuries. This climatic episode was characterized by colder temperatures, increased storminess, and significant advances of alpine glaciers beginning around 1450 and ending around 1850. Historical documents from Europe, including cod fishery and sea-ice reports from the seventeenth and eighteenth centuries, indicate sea-surface temperatures were 3°C to 5°C below the modern mean, ice floes penetrated south well beyond their normal extent, and the European coastline was repeatedly pummeled by torrential windstorms (Lamb 1982). Dozens of investigations from around the world of marine cores, sea-level curves, tree-ring chronologies, peat bogs, salt marshes, stalagmites, historic records, and even human tooth enamel have determined that the Little Ice Age was often synchronous across the globe throughout its four-hundred-year span (Soon and Baliunas 2003). Lean and Rind (1999) among many others point out that a decline in solar output was a primary forcing mechanism that triggered and sustained the Little Ice Age; they further suggest that an increase in solar output may equally be responsible for warming that occurred in the twentieth century, especially the first period of warming.

Solar Variability

Over the past few decades, scientists have determined that the sun is far from a constant star, and that its output varies at many time scales (Lean et

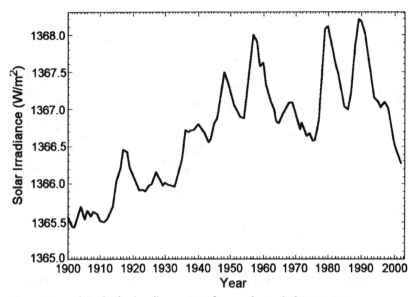

Figure 3.5.   Plot of solar irradiance (W/m²) over the period 1900–2002.

al. 1995; Lean 2001). Indeed, solar output has increased by approximately 2.0 Wm^{-2} over the period 1900 to 2002 (figure 3.5). Rather obviously, an increase in solar irradiance should translate into warmer earth temperatures. And there is a statistically significant positive relationship between solar irradiance and near-surface air temperature anomalies (figure 3.6). Over the period 1900–2002, solar irradiance values explain (statistically) nearly 40 percent of the variance in global annual temperature anomalies. However, when the data are broken into two subperiods, one from 1900 to 1969 and a second from 1970 to 2002, a differential pattern emerges. In the earlier subperiod, solar irradiance explains more than 50 percent of the variance in global temperatures with a sensitivity of 0.18°C per 1 Wm^{-2}. However, in the 1970–2002 subperiod, literally none of the variance in the global temperatures is explained by the solar irradiance values. The consensus view, as articulated by the most recent assessment of the IPCC, is that the global climate was strongly controlled by solar variability until approximately 1970, but sometime after that date, some climate forcing has overcome the solar–climate connection (Houghton et al. 2001); the buildup of greenhouse gases is the obvious candidate as the now-overriding thermal forcing of the climate system. Michaels et al. (2000), noting that the warming of dry, cold air (which is devoid of a natural greenhouse effect) in recent decades increases with the amount of that cold air, providing a dispositive "proof" of its

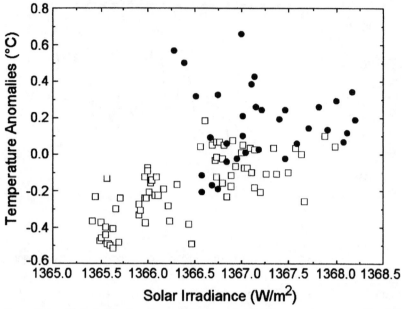

Figure 3.6.   Plot of solar irradiance (W/m²) and global temperature anomalies (°C) from 1900–1969 (open squares) and 1970–2002 (filled circles).

greenhouse nature. Nonetheless, if we control for solar variability over the 1900–2002 period, fully 42 percent of the warming over that period is accounted for in a statistical sense.

Despite the obvious physical linkage between solar output and the earth's temperature, scientists find themselves in a considerable debate about how small variations in incoming radiation elicit relatively large variations in global temperature. Most numerical models of climate fail to reproduce the statistical association revealed in figure 3.6. There must be a positive feedback mechanism that enhances the temperature response to the small changes in radiation.

One candidate for the enhancing feedback involves cosmic rays that may alter cloudiness in the upper or lower levels of the troposphere (Pallé and Butler 2002). Dickinson (1975) was among the first to explicitly suggest that "variations in cloudiness are likely to be related to variations in production of ionization near the tropopause by galactic cosmic rays." Basically, during times of high sunspot numbers, the earth's magnetic field weakens and cosmic rays from the sun and from outside the solar system penetrate into the troposphere. A series of complex microphysical processes may then take place that may stimulate the growth of high cloudiness. The cosmic ray feedbacks may help explain the strong correlation known to exist between the solar sunspot cycle length and near-surface air temperatures (Friis-Christensen and Lassen 1991). Many other mechanisms may be at work creating a positive feedback between solar output and planetary temperature, and this area of research should remain active for years to come.

Volcanic Eruptions

Some volcanic eruptions pump enormous amounts of dust into the stratosphere that can remain suspended for several years. This stratospheric dust has the effect of blocking out incoming radiation from the sun, thereby cooling the temperatures near the earth's surface. Many indices have been constructed to show variations in the effect of volcanism on the earth's radiation balance and temperature, but those indices are not well correlated among themselves (Robock and Free 1995). Nonetheless, many scientists have shown that a combination of solar variability with periodic volcanic eruptions explains substantial amounts of the variance in hemispheric and global near surface air temperature up to 1970, but not thereafter (Houghton et al. 2001). Once again, the warming from 1970 to present is difficult to reconcile with only known variations in solar and volcanic activity.

El Niño and La Niña

A significant, but slight, control of the planetary temperature comes from a major oscillation in atmospheric and oceanic circulation in the tropical Pacific (Ropelewski and Jones 1987). During some periods, the atmospheric circulation is invigorated, and the strong easterly flow off the equatorial coast of South America promotes upwelling in the ocean and relatively cold sea-surface temperatures. This phase of the "Southern Oscillation" is commonly referred to as La Niña, and its cold sea-surface temperatures over a substantial portion of the equatorial Pacific produce relatively cool global temperatures (figure 3.7). Oppositely, there are times when the easterly component of the tropical Pacific circulation is reduced, the upwelling weakens, and relatively warm water moves eastward across the Pacific leading to a large pool of warm water in the region. This condition is the ever-popularized El Niño event that tends to elevate global temperatures (figure 3.7).

The physical and dynamical linkage between the atmospheric and oceanic circulations makes it difficult to control for the Southern Oscillation

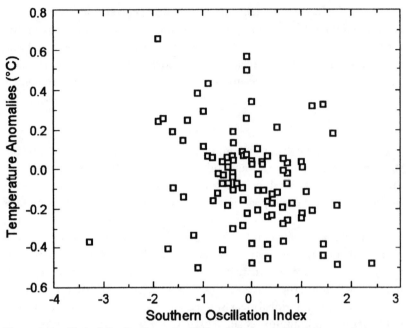

Figure 3.7.   Plot of the Southern Oscillation Index and global temperature anomalies over the period 1900–2002; negative values of the SOI indicate El Niño warm periods and positive values indicate La Niña cool periods.

in any attempt to isolate a global warming signal in the near-surface air temperature records. What is obvious, however, is that should the future see an increase in El Niño events that could be a consequence of the buildup of greenhouse gases (Meehl and Washington 1996), the global temperature will be forced to higher levels than expected.

Sulfate Aerosols

The first major scientific assessment of the IPCC in 1990 (Houghton et al. 1990) focused largely on the climate impacts of elevated concentrations of greenhouse gases, but two years later, the scientific assessment was revised to account for the cooling effects of elevated sulfur dioxide (SO_2) levels (Houghton et al. 1992). It is well known that burning fossil fuels produces not only CO_2 but also large amounts of SO_2. The SO_2 enters the atmosphere and quickly becomes an aerosol capable of reflecting incoming radiation, making clouds last longer, and brightening clouds, all of which have a cooling effect on the planet (Charlson et al. 1992). Given the short lifetime of sulfate aerosols in the atmosphere, however, the SO_2 disruption of planetary temperature should be highly regional. Given that the Northern Hemisphere has a much higher averaged atmospheric sulfate concentration than the Southern Hemisphere, more warming should therefore be evident in the Southern Hemisphere than the Northern Hemisphere over the past one hundred years. Yet the near-monthly averaged near-surface air temperatures for the two hemispheres are highly correlated, with no indication of the Southern Hemisphere warming faster than the Northern Hemisphere (figures 3.8, 3.9). While there is little doubt that SO_2 is having an effect on air temperature trends at regional scales, the effect on hemispheric temperature has been difficult to isolate.

Ozone

Ozone represents a powerful greenhouse gas, one that, in general, has been depleted somewhat in the stratosphere, particularly in the Southern Hemisphere's polar region. A series of papers appeared in the scientific literature in recent years suggesting that any depletion in stratospheric ozone should result in a cooling of the earth in the absence of any other changes in the thermal forcing of climate (Hansen et al. 1997, 1998). Conversely, near the surface, ozone concentrations have increased in urban areas, and the result should therefore produce a warming near the world's largest metropolitan areas.

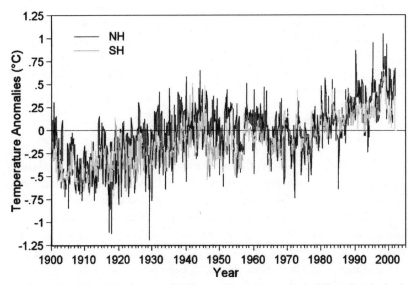

Figure 3.8.   Plot of northern (solid line) and southern (dashed line) hemispheric temperature anomalies during the period 1900–2002.

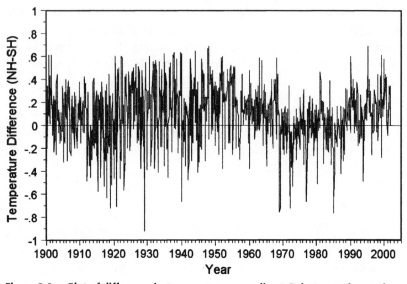

Figure 3.9.   Plot of difference in temperature anomalies (°C) between the northern and southern hemispheric temperature anomalies during the period 1900–2002.

Mineral Aerosols

The degradation of drylands has produced an increase in mineral aerosols (aka dust) at regional, hemispheric, and global scales. This dust load in the troposphere alters the transfer of radiation in the atmosphere and may result in a cooling effect. Tegen et al. (1996) conducted numerical experiments on the effects of mineral aerosols on air temperatures and concluded that the increase in atmospheric dust levels should have a net cooling at and near the earth's surface.

Other Forcings

The IPCC's 1990 scientific assessment dealt largely with the climate impacts of the ongoing buildup of greenhouse gases. By 1995, their scientific assessment was suggesting that the climate system is being perturbed largely by trends in greenhouse gases, stratospheric and tropospheric ozone, sulfates, fossil fuel soot, biomass burning, mineral aerosols, and solar variability. By 2000, however, the list had grown to include the additional potential effects of aviation-induced contrails and cirrus clouds as well as land-use changes. As time went on in the 1990s, the list of candidates potentially altering the climate system was growing faster than the buildup of greenhouse gases, and the uncertainties of where the climate was heading increased. Hansen et al. (1998) clearly stated that "the forcings that drive long-term climate change are not known with an accuracy sufficient to define future climate change." Even if we had accurate temperature data from throughout the world and highly accurate numerical models of climate, we would still be unable to forecast future near-surface air temperature levels given the uncertainties regarding the forcings of climate in the next fifty to one hundred years.

CONCLUSION

The global near-surface air temperature record is often used to support the claim that the earth is warming at a rate that is reasonably consistent with the popularized prediction for warming given the buildup of greenhouse gases. No matter how many compounding variables are brought to the table, many argue that the rise in greenhouse gases is or will soon be dominating all other radiative forcings of climate and will drive the global temperature further upward (Houghton et al. 2001). As the near-surface air temperatures continue to rise, some scientists will inevitably claim that the empirical evidence is highly supportive of that theory. And when observations are consistent with that theory, the theory gains considerable support from those who have a stake in promoting global warming fears.

There is little doubt that the global air temperature record has become a supporting ally in the ongoing greenhouse debate. Confounding many rather glib assertions is the fact that the warming rate in the early twentieth century (1915–1945) is not significantly different from the warming rate in the past three decades. However, the inordinate concentration of recent warming in very cold and dry air argues strongly for a greenhouse association.

But as this chapter makes clear, major problems remain. First, the temperature records are far from perfect and contain contaminations from urbanization, distribution of measurement stations, instrument changes, time of observation biases, assorted problems in measuring near-surface temperatures in ocean areas, and on and on. This could introduce a total bias of 0.2°C to 0.3°C, or about one-third of the observed warming. Second, even if we accept that the warming is real, there is a strong argument that approximately half of the warming—the portion that took place in the early twentieth century, was anything but a natural recovery from the Little Ice Age. Third, there are many non–greenhouse thermal forcings at work on the near-surface air temperatures at regional, hemispheric, and global scales, and it is very difficult to isolate the signal related to the buildup of greenhouse gases.

Nonetheless, the evidence is overwhelming that the near-surface air temperatures have increased in the past three decades. However, the greenhouse debate involves far more than what happens at the surface. The same models that predict warming near the surface also predict even more warming *above* the surface, yet the lower troposphere does not appear to be warming at a rate consistent with the models, as shown in the next chapter. Satellite-based and weather balloon-based measurements of lower-atmosphere temperature are not consistent with surface trends, and the differential pattern is not expected given our understanding of atmospheric processes. The National Research Council's 2000 report (Wallace et al. 2000) acknowledged that "if global warming is caused by the buildup of greenhouse gases in the atmosphere, it should be evident not only at the earth's surface, but also in the lower to midtroposphere." Warming near the surface with little to no warming in the lower to midtroposphere is *not* a clear greenhouse signal!

Finally, there is the inevitable discussion of policies aimed at slowing down or halting the warming at the surface, no matter what the cause. The language of the Kyoto Protocol is hardly aimed at the temperature of the lower to midtroposphere, but rather the temperature near the earth's surface. Many scientists have considered the effect of Kyoto Protocol on near-surface temperature trends, and as Wigley (1998), and many others have concluded, the effect of the Kyoto Protocol, even with all signatories' full compliance, would be undetectable above the natural noise of climate variability for the next century.

The history of the earth is marked by significant fluctuations in global temperatures. The modern science of climatology was concerned about global cooling in the 1970s and global warming since the late 1980s. As time goes forward, we will undoubtedly assemble more accurate records of near-surface air temperatures and their resultant trends and a greater understanding of the causal mechanisms of the variations and trends. Furthermore, we will be able to make more accurate assessments of how various policy actions will impact global temperatures. But at this moment in time we know only that:

- Global near-surface air temperatures have risen in recent decades.
- Lower-tropospheric temperatures have warmed little over the same period.

- This differential pattern is not consistent with predictions from numerical models.
- Proposed policy actions will produce trivial impacts on climate, despite their non-trivial negative economic consequences.

This era should be a time of scientific assessment, not a time of impulsive policy actions targeted at a problem that may not even be a problem, and will not be significantly impacted by those very policy actions.

REFERENCES

Arrhenius, S. 1896. On the influence of carbonic acid in the air upon the temperature of the ground. *Philos Mag* 41:237–76.

Balling, R. C., Jr., and C. D. Idso. 2002. Analysis of adjustments to the United States Historical Climatology Network (USHCN) temperature database. *Geophys Res Lett* 29, 25:1–3; doi:10.1029/2002GL014825.

Balling, R. C., Jr., and S. B. Idso. 1989. Historical temperature trends in the United States and the effect of urban population growth. *J Geophys Res* 94:3359–63.

Charlson, R. J., S. E. Schwartz, J. M. Hales, R. D. Cess, J. A. Coakley Jr., J. E. Hansen, and D. J. Hofmann. 1992. Climate forcing by anthropogenic aerosols. *Science* 255:423–30.

Christy, J. R., D. E. Parker, S. J. Brown, I. Macadam, M. Stendel, and W. B. Norris. 2001. Differential trends in tropical sea surface and atmospheric temperatures since 1979. *Geophys Res Lett* 28:183–86.

Dickinson, R. 1975. Solar variability and the lower atmosphere. *Bull Amer Meteor Soc* 56:1240–48.

Easterling, D. R., B. Horton, P. D. Jones, T. C. Peterson, T. R. Karl, D. E. Parker, M. J. Salinger, V. Razuvayev, N. Plummer, P. Jamason, and C. K. Folland. 1997. Maximum and minimum temperature trends for the globe. *Science* 277:364–67.

Friis-Christensen, E., and K. Lassen. 1991. Length of the solar cycle: An indicator of solar activity closely associated with climate. *Science* 254:698–700.

Gall, R., K. Young, R. Schotland, and J. Schmitz. 1992. The recent maximum temperature anomalies in Tucson: Are they real or an instrumental problem? *J Climate* 5:657–65.

Gallo, K. P., T. W. Owen, D. R. Easterling, and P. F. Jamason. 1999. Temperature trends of the U.S. Historical Climatology Network based on satellite-designated land use/land cover. *J Climate* 12:1344–48.

Grove, J. M. 1988. *The Little Ice Age.* London: Methuen.

Hansen, J., and S. Lebedeff. 1987. Global trends of measured surface air temperature. *J Geophys Res* 25:13345–72.

Hansen, J., M. Sato, and R. Ruedy. 1997. Radiative forcing and climate response. *J of Geophys Res* 102:6831–64.

Hansen, J. E., M. Sato, A. Lacis, R. Ruedy, I. Tegen, and E. Matthews. 1998. Climate forcings in the industrial era. *Proc of the Nat Acad of Sciences* 95:12753–58.

Houghton, J. T., B. A. Callander, and S. K. Varney (Eds.). 1992. *Climate change 1992: The supplementary report to the IPCC scientific assessment.* Cambridge: Cambridge University Press.

Houghton, J. T., G. J. Jenkins, and J. J. Ephraums (Eds.). 1990. *Climate change: The IPCC scientific assessment.* Cambridge: Cambridge University Press.

Houghton, J. T., L. G. Meira Filho, B. A. Callander, N. Harris, A. Kattenberg, and K. Maskell (Eds.). 1996. *Climate change 1995: The science of climate change.* Cambridge: Cambridge University Press.

Houghton, J. T., Y. Ding, D. J. Griggs, M. Noguer, P. J. van der Linden, X. Dai, K. Maskell, and C. A. Johnson (Eds.). 2001. *Climate change 2001: The scientific basis.* Cambridge: Cambridge University Press.

Howard, L. 1833. *The Climate of London.* London: Harvey & Darton.

Jones, P. D. 1994. Hemispheric surface air temperature variations: A reanalysis and an update to 1993. *J Climate* 7:1794–1802.

Karl, T. R., H. F. Diaz, and G. Kukla, 1988. Urbanization: Its detection and effect in the United States climatic record. *J Climate* 1:1099–1123.

Karl, T. R., R. W. Knight, and J. R. Christy, 1994. Global and hemispheric temperature trends: Uncertainties related to inadequate spatial sampling. *J Climate* 7: 1144–63.

Lamb, H. H. 1982. *Climate, history and the modern world.* London: Methuen.

Lean, J. L. 2001. Solar irradiance and climate forcing in the near future. *Geophys Res Lett* 28:4119–22.

Lean, J., J. Beer, and R. Bradley. 1995. Reconstruction of solar irradiance since 1610: Implications for climate change. *Geophys Res Lett* 22:3195–98.

Lean, J., and D. Rind. 1999. Evaluating sun–climate relationships since the little ice age. *J Atmos Solar-Terr Physics* 61:25–36.

Mann, M. E., R. S. Bradley, and M. K. Hughes. 1998. Global-scale temperature patterns and climate forcing over the past six centuries. *Nature* 392:779–87.

———. 1999. Northern hemisphere temperatures during the past millennium: Inferences, uncertainties, and limitations. *Geophys Res Lett* 26:759–62.

Meehl, G. A., and W. M. Washington. 1996. El Niño–like climate change in a model with increased atmospheric CO_2 concentrations. *Nature* 382:56–60.

Michaels, P. J., P. C. Knappenberger, R. C. Balling Jr., and R. E. Davis. 2000. Observed warming in cold anticyclones. *Clim Res* 14:106.

Nasrallah, H. A., and R. C. Balling Jr. 1994. The effect of overgrazing on historical temperature trends. *Ag For Meteor* 71:425–30.

Pallé, E., and C. J. Butler. 2002. The proposed connection between clouds and cosmic rays: Cloud behaviour during the past 50–120 years. *J Atmos Solar-Terr Physics* 64:327–37.

Parker, D. E., C. K. Folland, and M. Jackson. 1995. Marine surface temperature: Observed variation and data requirements. *Clim Change* 31:559–600.

Peterson, T., K. Gallo, J. Lawrimore, T. Owen, A. Huang, and D. McKittrick. 1999. Global rural temperature trends. *Geophys Res Lett* 26:329–32.

Rind, D., R. Goldberg, and R. Ruedy. 1989. Change in climate variability in the twenty-first century. *Clim Change* 14:5–37.

Robock, A., and M. P. Free. 1995. Ice cores as an index of global volcanism from 1850 to the present. *J Geophys Res* 100:11, 549–67.

Ropelewski, C. F., and P. D. Jones. 1987. An extension of the Tahiti–Darwin Southern Oscillation Index. *Monthly Weather Review* 115:2161–65.

Soon, W., and S. Baliunas. 2003. Proxy climatic and environmental changes of the past 1,000 years. *Clim Res* 23:89–110.

Tegen, I., A. A. Lacis, and I. Fung. 1996. The influence on climate forcing of mineral aerosols from disturbed soils. *Nature* 380:419–22.

Vinnikov, K. Y., P. Y. Groisman, and K. M. Lugina. 1990. Empirical data on contemporary global climate changes (temperature and precipitation). *J Climate* 3:662–67.

Wallace, J. M., J. R. Christy, D. J. Gaffen, N. C. Grody, J. E. Hansen, D. E. Parker, T. C. Peterson, et al. 2000. *Reconciling observations of global temperature change.* Washington, D.C.: National Academy Press.

Wigley, T. M. L. 1998. The Kyoto Protocol: CO_2, CH_4 and climate implications. *Geophys Res Lett* 25:2285–88.

Williams, M. A. J., and R. C. Balling Jr. 1996. *Interactions of desertification and climate.* London: Arnold.

4

TEMPERATURE CHANGES
IN THE BULK ATMOSPHERE

Beyond the IPCC

John Christy

My assignment as a lead author of the U.N. Intergovernmental Panel on Climate Change (IPCC) 2001 report was to draft the section on the temperature of the air above the surface. I must say at the outset that my personal experience of working on chapter 2, "Observed Climate Variability and Change," was quite positive. The process of producing such a document is tedious, involving the interaction of reviewers from around the world and the cooperation of people engaged in sometimes sharp disagreements. I specifically compliment the convening lead authors of chapter 2, Tom Karl (NOAA/NCDC) and Chris Folland (Hadley Centre, the U.K. Meteorological Office), for their tireless and careful leadership during the writing and editing of sections 2.2.3, "Temperature of the Upper Air," and 2.2.4, "How Do Surface and Upper-Air Temperature Variations Compare?" (Folland et al. 2001).

These two upper air sections were limited to about 2,000 words in the 40,000-word chapter, and were necessarily concise and clinical, with little opportunity for expanded comment. Over the three-year period in which the chapter was written (late 1998 to 2001), numerous scientists and government officials reviewed and commented on the drafts. Editing occurred up to the last minute, even after the opportunity for the primary authors to continue having influence had passed. In a parallel task, a shortened technical summary, about fifty pages long, was being written by a selected group of lead authors and other individuals even as the main text was being developed. An even shorter document (about twenty pages) known as the Summary for Policymakers, was simultaneously written in almost outline form. Those two summaries attracted the most attention, since their accessible digest format resulted in a level of scientific detail that the media found useful. As you might expect, however, condensing material from a text of eight hundred pages down to fifty or twenty is not easy, and controversies ensued as to why some climate change bytes were included and others left out. Why, for example, was a full page of the brief Summary for Policymakers devoted to surface temperature charts that depict considerable warming with another half page of supporting text, while changes in the full bulk temperature of the atmosphere—far more important for the physics of the greenhouse effect—garnered but seven sentences? Indeed, the development of the technical summary and Summary for Policymakers would make for an interesting investigation, but I shall leave that for others to do (see sidebar for a personal observation on the IPCC).

As with any endeavor of this type, some of the most difficult decisions relate to the "drop dead" cutoff date beyond which no newly published material may be included, no matter how pertinent it might be. For IPCC 2001, that date was actually as late as March 2001, though the IPCC provided the media with press releases based on almost finalized versions as early as 2000, and formally at the January 2001 IPCC meeting, where these near final drafts were officially approved. (The desire of the chapter 2 lead authors to gather and include data through the end of 2000 was the main cause of our delay.) My effort here will attempt to update and expand on the findings of IPCC 2001, as well as those of the National Research Council (NRC) (2000a) report *Reconciling Observations of Global Temperature Change*.

What follows will, at times, approach the level of technical discourse found in the IPCC itself and may be difficult to understand. That is especially true for the comparison between the two satellite-based atmospheric temperature data sets—crucial because of recent reports claiming one is better than the other, claims not based not on observations but on climate model output.

BACKGROUND

In the climate change arena, it is essential to recognize the importance of upper-air quantities. Although we live and function on the underlying surface of this massive atmosphere, we misrepresent the issue of climate change if we focus only on surface conditions and projections of surface climate. Obviously the atmosphere as a whole, whose mass is about 10,000 kg above each m^2 on the earth's surface, plays a significant and indeed dominant role

Although I enjoyed working on the IPCC, I found that most of the lead authors clearly favored the Kyoto Protocol. Several stated to me that the IPCC report should provide the necessary supporting evidence to persuade governments to adopt the treaty. No less than the head of the IPCC, Dr. Robert Watson, testified alongside me at congressional hearings, adamantly advocating acceptance of the Kyoto Protocol. Those hearings occurred during the development of the IPCC 2001 documents. Thus from the top of the IPCC organization on down, a bias was easily evident toward a specific policy action among many of the authors. If a poll had been taken, I would guess at least 80 percent of the lead authors would have supported Kyoto Protocol.

Given that situation, it is not too difficult to locate sections in IPCC 2001 that appear to withhold the scientific objectivity that could detract from a view that dangerous climate change was a certainty, as projected by the models. One example I found dealt with the diminishing sea ice extent in the Arctic. On pages 74 and 75 is a plot of two model projections of Arctic sea-ice extent along with observations (see figures 4.1 and 4.2). The pertinent part of the diagram is distilled into one plot, given below, in which the observed rate of decline in the latter part of the twentieth century is reasonably replicated by two coupled climate models. That prompted the IPCC authors to conclude:

> The simulations of ice extent decline over the past thirty years are in good agreement with the observations, lending confidence to the subsequent projections which show a substantial decrease in Arctic sea-ice cover.

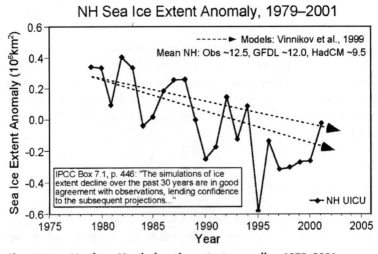

NH Sea Ice Extent Anomaly, 1979–2001

IPCC Box 7.1, p. 446: "The simulations of ice extent decline over the past 30 years are in good agreement with observations, lending confidence to the subsequent projections..."

Models: Vinnikov et al., 1999
Mean NH: Obs ~12.5, GFDL ~12.0, HadCM ~9.5

NH UICU

Figure 4.1.   Northern Hemisphere ice extent anomalies, 1979–2001.

(*continues*)

(*Continued*)

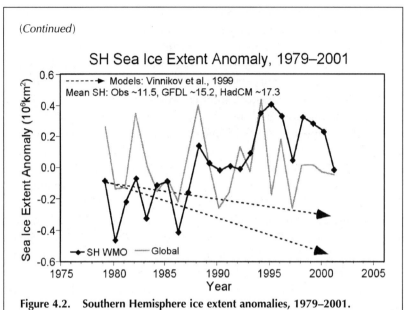

Figure 4.2. Southern Hemisphere ice extent anomalies, 1979–2001.

After reading that passage, I wondered whether a skeptical reporter or scientist was around to ask a simple question: How did the models do in the Southern Hemisphere? Evidently no such inquisitor was available, so I looked into it myself, admittedly long after the document was published. The necessary data were readily accessible through the kind intervention of K. Vinnikov. The SH result in the second plot tells a remarkably different story about the ability of coupled climate models to project reality. The dashed lines from the same two models are moving with a trend *opposite* that of the observations as available in 2000 (solid black line). Indeed the total global sea-ice extent (solid gray line) has no long-term trend. I'm wondering how the conclusions of this section in the IPCC 2001 would have changed if this "overlooked" data were displayed in the same large diagram as that which the IPCC used for the Arctic.

in controlling the potential human-induced climate changes that might ultimately impact the surface. In fact, the processes that transpire in the deep atmosphere are precisely those that produce the weather that affects the surface. The bulk atmosphere, with its 10,000 kg per m², is one key repository where energy trapped by the enhanced greenhouse effect must be stored, according to climate models. It is through the bulk temperature of the atmosphere that we can track such energy changes (Pielke 2003). Therefore, the processes and variations of the atmosphere in total must be investigated in any assessment of climate variability and change. As a corollary, the full virtual atmosphere represented in climate models must be evaluated against the data sets of upper-air quantities now being produced to test their ability to transfer energy properly throughout the climate system.

Physically, temperature is an upper-air variable of state that, according to the laws of physics, will respond rapidly to increases in infrared absorbing gases, all things being equal. In other words, additional energy trapped in the system will reveal itself by increasing the bulk atmospheric temperature. As a consequence, it may be one of the earliest and clearest indicators of human-induced climate change (Barnett 1986). Secondary effects, which could, for example, be manifest in changes in storminess, are related to changes in the vertical and horizontal distribution of temperature (or thermal energy) and thus could also be indicators of the influence of greenhouse gas increases, though with less certainty. The rate at which radiation is distributed and exhausted to space is tied to the temperature structure of the atmosphere, thus alterations in the vertical structure have implications for the radiational balance of the earth system. Simply put, the upper-air temperature, as a direct indicator of energy content, provides key information on essentially all aspects of climate and climate change.

MEASURING UPPER-AIR TEMPERATURES

Several monitoring systems have been utilized in the past seventy years to provide information of the atmosphere above the surface. (Further information is given in the IPCC reports of 1996 and 2001.) Initial requirements for upper-air measurements arose with the aviation era, as pilots discovered unexpected winds that at times maintained velocities greater than an aircraft's typical ground speed. As time went on, pilots in World War II needed flight level information along new military flight lines. Some form of systematic measurement of the upper atmosphere was therefore required. (Note that the measuring systems arose out of the needs of aviation, not climate science.)

Weather Balloons

The first data sets to provide long-term upper-air weather information came from instrument packages carried aloft by balloons filled with helium (or hydrogen in the early years and also in the most recent years due to helium's rising cost), which were capable of transmitting data to a ground station nearby via an instrument called a radiosonde. Early packages were crude, with the main quantities of interest being the speed and direction of winds, determined by visual or radar tracking as the balloon ascended. Temperature, pressure, and humidity were more difficult to measure accurately, as numerous factors influenced the readings. For example, the thermistor (for temperature) always experiences a time lag in its response to the temperature of the air through which it ascends, making it difficult to assign a precise vertical level to the temperature reported. Another complicating phenomenon is that the sun would heat the package's housing or sensor itself, which tended to produce temperatures warmer than that of the ambient air.

These time-lag and solar-heating problems are basically unique for each type of instrument employed at the upper-air stations. As a result, creating a homogeneous time series of upper-air quantities is a nontrivial task. To multiply the uncertainties further, many countries build their own versions of radiosondes, and new versions have been continually developed and used through the years. The lack of systematic instrumentation through time combined with virtually no on-site intercomparison between old and new devices created discontinuities that in many cases were found to be significant (Gaffen 1994; Parker and Cox 1995). The precision required for assessing the longer-term changes so important to climate studies was not a requirement with those instruments. For example, a change from one instrument to another of a different manufacturer could produce a discontinuity of 3°C at the lower stratospheric levels (Parker et al. 1997), which completely compromises the ability to detect trends on the order of 0.05°C decade^{-1}.

One of the first efforts to construct a global data set of upper-air temperatures is documented by Angell (1988, 2000), who assembled data from sixty-three radiosonde stations with mostly complete records and with few instrument changes.[1] Angell calculated the thickness temperature between distinct pressure levels (e.g., 850–300, 100 to 50 hPa, etc.) (see table 4.1 for the relationship between pressure levels and altitude) as the quantity of interest. Seasonal temperature data from these sites were averaged into seven latitudinal bands from which global averages were generated. Among this data set's significant contributions to climate research: evidence for large interannual variations such as sudden winter stratospheric warming near the pole and volcanically induced tropospheric cooling with stratospheric warming.

Questions regarding the veracity of the long-term trends in these unadjusted, composited data led to further studies. Angell determined that at least nine stations experienced significant shifts toward cooler temperatures primarily in the higher levels, which would lead to spurious negative trends

Table 4.1.   Atmospheric Pressure and Altitude. Atmospheric scientists traditionally assess the vertical profile of the atmosphere at a level where barometric pressure is constant. This is determined by the height to which a measuring instrument must ascend to record such a pressure. The pressure unit is the hectoPascal (hPa), which is also the weight of the atmosphere (in millibars) remaining above the barometer. The relationship between pressure (hPa) and altitude above sea level (feet) is given in the U.S. Standard Atmosphere. Note that the altitude in a "standard atmosphere" is an average. In reality, the 850 hPa height is lower near the poles (where the cold atmosphere is more dense) and higher in the tropics. In fact, it is the geographic distribution of these pressure-determined altitudes that defines temperature changes in the troposphere.

Altitude (ft)	Pressure (hPa)
0	1013.2
4,800	850
9,600	700
18,300	500
30,000	300
38,500	200
53,200	100

in the 1000–100 hPa and 100–50 hPa temperatures (Angell 2002, personal communication). Gaffen and colleagues (2000a) attempted to quantify possible discontinuities in many of those same stations through statistical tests that identified change points—sudden temperature shifts not likely to be natural. These researchers obtained partial information on changes in practices and instrumentation (i.e., metadata) at many of the sites; however, given the relatively large natural variability of a single station's temperature series (generally having greater variance than surface temperatures) and lack of complete metadata records, the Gaffen team found they could produce different sets of "corrected" time series that displayed significantly different trends for the same station. The differences were due solely to varying the predetermined set of parameters designed to locate and correct for change points. Making the situation more difficult was that for the period 1959–1995, only twenty-two stations of those examined contained enough data to establish trends up to the lower stratosphere (30 hPa).

From this research, a team from NOAA set out to create an adjusted data set of pressure-level temperatures for eighty-seven stations distributed fairly evenly throughout the globe and with a minimum of data gaps. The procedure involved objective information known as metadata, which is the history of instrumental or procedural changes (see Gaffen 1993); statistical indications of change points; and subjective insight as to events in the time series that appeared to be unnatural. The group members would individually develop recommended changes, then meet together and by consensus create and apply appropriate adjustments (Lanzante et al. 2003a, b). The first version of this data set, known as LKS (for Lanzante, Klein, and Seidel), contained information through 1997. Further versions will utilize an automated adjustment system for post-1997 data.

Scientists at the U.K. Meteorological Office's Hadley Centre created several versions of radiosonde-based temperature data sets. These Hadley Radiosonde Temperature (HadRT) data were derived from the CLIMAT TEMP reports of monthly mean temperatures by about four hundred station operators at the "mandatory" pressure levels. As an additional set of products, the Met Office produced simulated satellite temperatures from radiosonde data for more direct radiosonde versus satellite comparisons (Parker et al. 1997) The data were binned into 5° latitude by 10° longitude grid boxes, with some minor infilling of missing grids where sufficient adjacent grids with data were available for interpolation.

The U.K. researchers came across many situations in which the data were obviously discontinuous. To reduce the influence of those discontinuities, Parker et al. identified dates in each station's time series for which a change at the data stream was documented, and then estimated various types of corrections based on comparisons across the identified discontinuity with UAH (University of Alabama in Huntsville) microwave-based temperatures as measured by satellites (but only since 1979; see below). From this came, for example, version HadRT2.1s, the main data set utilized here. That version contained data whose stratospheric levels only were adjusted at the dates specified to agree with the UAH MSU/AMSU data.[2] Version HadRT2.0 represented uncorrected, raw data. Other versions included tropospheric corrections based on UAH MSU/AMSU (e.g., HadRT2.2) and globally complete data in which the NOAA NCEP Reanalysis (see below) was used to fill in areas for which no radiosonde stations existed (HadRT2.3). In general, tropospheric adjustments were minor and had little impact on decadal trends while stratospheric adjustments were more significant.

A new Hadley Centre data set, HadAT, is being developed that does not use other types of data for corrections, but relies on the horizontal spatial coherency of pressure-level temperature anomalies for determining the impact of the documented changes in the station data (Thorne et al. 2004). As with LKS, this HadAT version will be completely independent of the satellite data sets.

The Russian Research Institute for HydroMeteorological Information (RIIIMI) generates monthly global gridded temperature data using more than eight hundred radiosonde stations on the mandatory pressure levels (Sterin 1999). Those grids without direct observations are filled in by interpolation between radiosondes, which requires interpolation across large distances, for example, in the Southern oceans. With no local information in such areas, the overall variance of global average anomalies is significantly reduced (Christy et al. 2003a). Note that layer temperatures of NOAA Angell and RIHMI do not match the weighting profiles of the MSU/AMSU data sets, so some discrepancy should be expected. This is especially true in the lower stratosphere (100–50 hPa) where the lower stratosphere (LS or MSU4/AMSU9) profile includes more than 20 percent of its weight below 100 hPa, which is in the tropical troposphere, a layer that has experienced warming in the past few decades. One should expect that trends should be more positive in the LS profile versus the 100–50 layer temperature as a result.

Temperatures Sensed by Orbiting Microwave Radiometers

Since late 1978, NOAA polar orbiting satellites have carried a passive microwave radiometer to monitor emissions of atmospheric oxygen at several frequencies near the 60 GHz absorption band. Each frequency receives emissions from a broad layer of the atmosphere, represented by a weighting function. For example, channel 2 of the Microwave Sounding Unit (MSU2) receives emissions from the surface to the stratosphere, with a peak signal in the midtroposphere (approximately 15,000 feet) tapering off above and below. The products of interest here are the temperatures of the lower troposphere (LT, surface to about 8 km), midtroposphere (MT, surface to about 18 km) and lower stratosphere (LS, 15 km to 23 km). LT is produced as a linear combination of view angles of MSU2; MT is MSU2; and LS is derived from MSU4. With the launch and commissioning of the NOAA-15 spacecraft in August 1998, the MSU was replaced by the Advanced MSU (AMSU), incorporating several more channels. MT is represented by

AMSU5, LS by AMSU9, and LT by a linear combination of AMSU5 view angles (Christy et al. 2003a).

Passive microwave emissions lend themselves to climate products since the signal is largely dependent on atmospheric temperature alone. The small impacts of clouds, dust, water vapor (heavy rain is screened out) and inter-annual variations in surface emissivity are small and thus have virtually no impact on the long time series (Spencer and Christy 1990). But other factors that create spurious trends in the time series of the same magnitude as the signal being sought must be dealt with. For example, the "sun-synchronous" polar orbiting spacecraft tend to drift from their initial crossing time to later or earlier in the day over a period of years. A 1400 (2:00 P.M.) crossing time may drift to 1700 (5:00 P.M.), or late afternoon local time, which introduces a spurious cooling trend because the atmosphere naturally cools between 1400 and 1700 in its diurnal cycle (Christy et al. 1995). Instrument calibration issues also have been discovered (Christy et al. 2000), as well as satellite biases (Spencer and Christy 1990) and orbital decay effects (Wentz and Schabel 1998).

The University of Alabama in Huntsville (UAH) produces time series of LT, MT, and LS that merge data from the nine MSUs and two AMSUs. Generally only two spacecraft are operational at any given time, each lasting about four to six years, each launched to replace an older one that has failed or is nearing retirement.

Remote Sensing Systems (RSS) of San Rafael, California, has recently begun producing MT and LS products from raw MSU/AMSU data using methodologies that differ in some respects from those of UAH to account for the required adjustments (Mears et al. 2003.) Those differences are addressed below.

Details of the Microwave-Based Products

There are now two globally gridded microwave-based data sets with somewhat different results for one product (MT). For that reason, it is necessary to delve into some detail to search for an explanation of the differences. The reader who is interested in the influence those differences have on outcome may wish to skip this particularly technical section and proceed to the section "Error Estimates."

INSTRUMENT BODY EFFECT

Christy et al. (2000) reported the discovery of the instrument calibration problem and developed a technique to account for it. This problem was found when the global mean temperatures of two coorbiting satellites were compared and systematically varying differences appeared. These differences in most cases were significantly correlated with the temperature of the instruments themselves as monitored by the platinum resistance thermometers embedded in the hot-target plate. A technique was developed to calculate a coefficient for each spacecraft which when multiplied by its time-varying, hot-target temperature would produce a correction and reduce the intersatellite differences. In essence, for two individual spacecraft (1, 2) the coefficients (a_1, a_2) were determined as solutions to a large system of linear equations which related the daily hot-target temperatures(T_{HOT1}, T_{HOT2}), their daily global temperatures (T_{GL1}, T_{GL2}) and their systematic bias $(T_{BIAS1v2})$, that is,

$$a_1 T_{HOT1} - a_2 T_{HOT2} = T_{GL1} - T_{GL2} + T_{BIAS1v2} + \text{error.}$$

T_{BIAS} and the two coefficients a_1 and a_2 were determined by regression based on the reduction of the residual error to its minimum.

In the course of developing this correction in 1998, UAH determined that thresholds on the method were preferred since in some cases the magnitude of the reduction of error was meaningless (i.e., less than 10 percent). As a result, UAH used intersatellite differences from coorbiting satellites only if the period length was a minimum of one year. In addition, and not unrelated, the resulting reduction of variance was required to exceed a threshold of 40 percent for LT and MT and 25 percent for LS.

Finally, the greatest reduction of error variance, especially in terms of reducing trend differences between the satellites, was achieved by first smoothing the daily data before proceeding with generating the large system of equations (Christy et al. 2000).

RSS extended the technique of UAH to include all of the overlapping pairs, regardless of length or variance explained. In addition, the unit of time in the RSS data set was pentads (five days), with no smoothing when calculating the target coefficients. Thus fundamental differences between UAH and RSS methods relate to the use of (1) thresholds and (2) smoothing by UAH. UAH focused on reducing longer-frequency intersatellite differences (i.e., intersatellite trend differences) while limiting the system of equations to those with the most reproducible data. RSS concentrated on reduction of error variance of the higher frequencies for every overlap available, thus utilizing as much information as possible.

The resulting target coefficients are similar between UAH and RSS except in the case of NOAA-9, the fifth satellite in the series. The difference in NOAA-9 MSU2 target coefficients (UAH: -0.10 vs. RSS: -0.04) arises from the UAH smoothing procedure and whether the other shorter overlaps are employed in solving the large system of linear equations (for details, see Christy et al. 2000; Mears et al. 2003). The magnitude of virtually all other target coefficients for both MT and LS are within ±0.04, raising concern that a UAH value of -0.10 is unusually large in magnitude (Mears et al. 2003).

The difference between UAH and RSS in MT temperatures related to this effect is most prominent in the region south of 20°S. In the tropics, there is no difference between the two products in the trend of MT until after the NOAA-9 period, implying other factors are responsible for trend differences there. Globally, the high frequency error is reduced best by RSS (daily standard error 0.037°C vs. 0.049°C for the overlap between NOAA-6 vs. NOAA-9), while the trend difference between the two is best reduced by UAH (0.04 vs. 0.01°C year^{-1}). The impact of this differing MT target coefficient on the global trend is $+0.05$°C decade^{-1} when applied to UAH data. The 1979–2003 trends of MT for UAH and RSS are $+0.044$ and $+0.124$°C decade^{-1} respectively. Thus the difference in the target coefficient for NOAA-9 explains about 60 percent of the overall difference in global trends. In comparison with independent radiosonde data, there are no significant departures between UAH and the radiosondes during the NOAA-9 period of influence (Christy and Norris 2004). After the NOAA-9 period (post 1986), the two data sets have relatively small differences.

DIURNAL DRIFT EFFECT

A second major difference between UAH and RSS methodologies relates to corrections required to account for the slow east–west drift of the satellites during their multiyear periods of service. For example, to calculate the spurious cooling induced as a satellite drifts from 2:00 P.M. to 5:00 P.M. over a fixed spot, UAH developed a correction table derived from the satellite observations themselves. The MSU and AMSU are cross-track scanning instruments, meaning the instrument sweeps left to right (west to east) when on its northbound pass, observing several west–east oriented "footprints" during each few-second scan. The first or westernmost footprint represents the temperature at an earlier local time than the remaining footprints. As a result, the mean change in temperature of the cross-scan observations represents the mean change in temperature across the local times so observed when averaged over tens of thousands of cases.

A diurnal cycle was constructed from these observations for land and ocean separately, each month individually and each separate latitude. Thus, UAH employs an empirical method of diurnal cycle corrections. A final point is that UAH developed corrections for each satellite to be applied from the satellites' initial time of insertion into the orbit—that is, the initial data were not adjusted to a common reference time in the diurnal cycle before further adjustments were applied as in the RSS method.

RSS obtained the hourly output of the NCAR CCM3 climate model over a mean annual cycle. From this, RSS determined for each grid a diurnal cycle based on simulating the model-derived satellite temperature that would be observed at each hour. The advantage of this technique is that for a given latitude band with more than one type of land surface (or more directly, more than one type of diurnal cycle over land), there would be a specific diurnal correction available. A major portion of the diurnal cycle in MSU2 is therefore introduced by the surface. In the UAH method, each latitude has a single, monthly dependent correction for all land in that band. (Note: The main products of UAH are generated from daily latitude-band averages, and so only require latitude-band corrections.)

A disadvantage of the RSS approach is that it is likely the CCM3 does not reproduce all of the fine structure of the diurnal cycle that is responsible for the drift errors. For example, the higher modes of the diurnal cycle, which lead to the diurnal drift "errors," are impacted by asymmetries in the daily cycle in convection (amplitude of about 14 percent), downward propagating stratospheric tides, direct atmospheric and cloud absorption of solar radiation, and so on. The asymmetry in the MT diurnal cycle is seen globally in that the mean minimum temperature beginning at 0630 rises until early afternoon (1400, or 8.5 hours later) then cools over the next 15.5 hours. Thus the diurnal cycle is complex. It is this difference in the diurnal cycle correction between UAH and RSS which may be responsible for the minor difference in trends in the post NOAA-9 (post 1987) period, seen most prominently in the tropics.

PATCHING THE INDIVIDUAL SATELLITE
SERIES INTO ONE TIME SERIES

A third difference in the merging techniques relates to the calculation of intersatellite biases. UAH applies all of the temporal adjustments to the zonal mean anomalies (e.g., target coefficient adjustments, diurnal drift adjustments, etc.) of each individual satellite, and as a last step calculates the intersatellite biases, latitude by latitude, by direct comparisons. This was done because it was demonstrated that there was a latitude dependence on the biases between various satellite instruments. In addition, there were also differences between the biases of this direct method and the biases (T_{BIAS}) that are produced as a byproduct of the linear system of equations that generate the target coefficients (Christy et al. 1998, 2000). This direct method was tested and found to be highly reproducible (Christy et al. 1998). It was also shown that the added error of the short overlapping periods was too great to include in building the basic time series. These biases are then removed and the entire set of satellite records is merged.

RSS uses the single bias, T_{BIAS} per satellite, applying it to all latitudes and then merges the satellite records. The RSS intersatellite biases are those consistent with the solution set of the linear system of equations from which the target coefficients are determined (Mears et al. 2003). Because RSS utilizes all overlapping periods in building their system of linear equations, the solution set will include biases for each satellite relative to all others with whom it shares an overlapping period. UAH produces and applies only those biases which are necessary to build the backbone of the data set (TIROS-N vs. NOAA-6, N-6 vs. N-9, N-9 vs. N-10, N-10 vs. N-11, N-11 vs. N-12, N-12 vs. N-14, N-14 vs. N-15, and N-15 vs. N-16) and connect all other time series to it. RSS incorporates the additional biases of the relatively short overlaps of NOAA-7 vs. N-8, N-8 vs. N-9, N-10 vs. N-12, and N-11 vs. N-14.

Christy et al. (1998) and Mears et al. (2003) reach different conclusions as to the value of the information in these shorter overlapping observations, though both show there is increased error relative to the longer overlaps associated with them. In other words, RSS solves an overspecified problem, and so reaches a statistical consensus of all of the data segments available. In this formulation, there will remain nonzero differences between various satellite pairs. However, because UAH defines a single, unique path based on the most reproducible results to create a backbone, all important intersatellite biases are eliminated.

Error Estimates

Christy et al. (2003a), through a series of comparisons with radiosonde data, conclude that the 95 percent confidence interval (C.I.) for the global trends of LT, MT, and LS are respectively, ±0.05, ±0.05, and ±0.10°C decade^{-1}. In a further study, Christy and Norris (2004) demonstrated that there was no difference between the composite trend of 101 balloons and collocated UAH LT values to within ±0.03 °C decade^{-1}, further bolstering the confidence in UAH techniques. It is important to remember that UAH and balloon data are completely independent.[3]

Mears et al. (2003) indicate for MT a 95 percent C.I. of ±0.02°C decade^{-1}, based on an error analysis of the target calibration matrix, though recently they upped the estimated error to about ±0.04. No error impacts of the climate-model diurnal cycle were applied, and, no comparisons with independent data were included, however. Given the MT global trend difference of 0.08°C decade^{-1} (UAH +0.044, RSS +0.124°C decade^{-1}), it is apparent that one or both of those error ranges is underestimated. In general, independent radiosonde-based trend calculations show greatest consistency with UAH rather than with RSS (see below).

Reanalyses

The European Centre for Medium Range Weather Forecasts (ECMWF) and the National Centers for Environmental Prediction (NCEP, USA, Kalnay et al. 1996) employ a fixed assimilation scheme within a global atmospheric model to take advantage of all possible data available for ingestion to produce global data sets of atmospheric quantities. The resulting products are consistent with model physics as the same model is used throughout the period of reanalyses. As the assimilation procedure is running, poor data, as determined by a number of factors, are screened out and replaced with a forecasted estimate or a value comparable with nearby acceptable data. But time-dependent biases may influence the Reanalyses and these are difficult to detect, especially in data sparse regions where the models are left on their own to compute the necessary quantities consistent with their internal physics. In addition, time-dependent biases, for example, from unadjusted satellite data, would appear as real trends in the output if not screened (Stendel et al. 2000; Trenberth et al. 2001).

The metric of greatest interest in this discussion is the long-term trend. In the NCEP data analysis system, the satellite temperature profiles (which include raw MSU data) needed to provide global coverage, are adjusted weekly by coincident radiosonde comparisons. Thus, long-term trends in NCEP are constrained by radiosonde data, not by the microwave satellite data described above. However, NCEP does show two substantial discontinuities. In the transition to incorporate satellite data in 1979, a significant jump in stratospheric temperatures was noted (Santer et al. 1999). Again, in March 1997 a significant drop in 100 hPa temperatures was discovered (Christy et al. 2003a). The first shift is simply associated with the availability of a new and massive source of data for the system. The cause of the second shift is less clear, but likely relates to a change in satellite data ingestion, because 100 hPa is generally the level at which radiosonde data begin to be discounted and satellite profiles accepted. As a result, trends that rely on data above 200 hPa (i.e., MT and LS) will be spuriously affected by the shifts in the NCEP time series. The trend values in table 4.2 are printed in italics to alert the reader to likely inaccurate trends.

Table 4.2. Temperature trends (°C/decade) over different time intervals from a variety of data sets and atmospheric levels. See text for details.

	1958-2002 Globe	Tropics	1979-1997 Globe	Tropics	1979-2002 Globe	Tropics
Surface						
HadCRU	+0.11	+0.09	+0.15	+0.13	+0.17	+0.12
GISS LSAT + HadISST	+0.11	+0.09	+0.11	+0.11	+0.15	+0.11
NCDC + NCEP SST	+0.11		+0.13		+0.16	
Lower Troposphere (LT)						
HadRT2.1s LT	+0.12	+0.13	-0.03	-0.12	+0.05	-0.05
NCEP LT[1]	*+0.15*	*+0.10*	-0.04	-0.08	+0.08	-0.03
UAH LT[2]			+0.00	-0.04	+0.07	-0.03
NOAA LKS LT			+0.06	+0.06		
NOAA Angell 850-300	+0.08	+0.07	-0.07	-0.11	-0.01	-0.10
Median	+0.10	+0.10	-0.03	-0.08	+0.06	-0.04
Mid-Troposphere MT						
HadRT2.1s MT	+0.04	+0.06	-0.13	-0.20	-0.06	-0.15
NCEP MT[1]	*+0.15*	*+0.10*	-0.11	-0.13	*-0.02*	*-0.09*
UAH MT			-0.05	-0.02	+0.03	+0.03
RSS MT			+0.05	+0.07	+0.11	+0.11
NOAA LKS MT			-0.05	-0.03		
NOAA Angell 1000-100[2]	*+0.02*	*+0.01*	*-0.15*	*-0.16*	*-0.07*	*-0.14*
Median			-0.05	-0.03	+0.03	+0.03
Lower Stratosphere						
HadRT2.1s LS	-0.40	-0.34	-0.66	-0.67	-0.66	-0.58
NCEP LS[1]	*-0.14*	*-0.05*	-0.71	-0.67	*-0.80*	*-0.64*
UAH LS			-0.53	-0.36	-0.49	-0.35
RSS LS			-0.39	-0.29	-0.37	-0.29
NOAA LKS LS			-0.77	-0.82		
NOAA Angell 100-50[3]	*-0.63*	*-0.69*	*-1.26*	*-1.46*	*-1.14*	*-1.28*
Median	-0.40	-0.34	-0.69	-0.52	-0.58	-0.47

Data likely subject to erroneous trends is in italics and not used in median determination (see text).
[1] Significant error in trends (italics) likely due to known shifts in time series (see text).
[2] Error ranges of the trends in the "tropics" column will be larger than those of the globe due to fewer observations.
[3] Significant negative error trends present due to unadjusted shifts in some stations.

Comparison of Available Data Sets

As indicated earlier, one of the key metrics of interest in understanding climate change and in dealing with controversies on upper-air temperatures is the trend. That metric is most sensitive to errors in the time series and so assessing decadal trends is a demanding test for these time series, especially when the differences between the data sets are generally less then $0.10°C$ decade^{-1}. Indeed the year-to-year variability is generally well captured by all of the data sets presented here so that intercorrelations are high (Seidel et al. 2004).

The issue of global warming deals precisely with the evidence for increased temperatures over decadal time scales that must be extracted from time series characterized by some level of error as well as significant interannual variability related to such events as ENSOs and volcanoes. Yet below the stratosphere the anticipated rate of human-induced warming is on the order of $0.1°C$ to $0.3°C$ decade^{-1} so that errors or interannual impacts of $0.1°C$ decade^{-1} approach the magnitude of the signal being sought. As a result, much attention is given to assessing trends and their errors by observationalists and modelers.

It should be noted that the satellite data sets discussed above cover a period of twenty-four years, which for many purposes is relatively brief for basing conclusions of attribution regarding climate change. Indeed there are two types of error to be kept in mind: (1) measurement error and (2) sampling error. The former deals with the errors due to instrument calibration, diurnal drift, and so on. Sampling error places the data in the context of time and space and might answer a question such as "How representative is the trend of this sample (1979–2002) relative to previous and future twenty-four-year periods?" Though we have fairly specific tests to estimate measurement error, it is more difficult to assess sampling error due to the apparent lack of representativeness of statistics over periods of only twenty-four years (Santer et al. 1999, 2001). For example, two significant volcanic events and two of the most intense El Niño–Southern Oscillation (ENSO) events have occurred only since 1979. Events of that magnitude did not occur in the previous forty years, thus the representativeness of the post-1979 period can be quite difficult to quantify (Christy et al. 2000; NRC 2000a; Folland et al. 2001).

We shall mainly focus on global and tropical variations and trends in the troposphere.[4] In the tropical troposphere, the vertical transport of heat is dominated by convection and subsidence through direct circulations. In virtually all climate model simulations, an effect of enhanced greenhouse gas increases includes a warming of the troposphere at a rate faster than the surface (i.e., a lessening of the lapse rate). With several observational estimates of the tropospheric and surface rates of warming (trends) we will be in a position to test this model result. In addition, bulk atmospheric temperatures are closely related to the energy content of the atmospheric system—the number of joules contained in the atmosphere. By keeping track of the energy content in the observed atmosphere and in modeled atmospheres, we may discover where difficulties in the heat exchange system of a model might appear.

The observational data are displayed in figure 4.3 and table 4.2. The significant tropospheric interannual variability is evident as seen by warm phase ENSO events (e.g., the 1998 El Niño), cool phase events (e.g., the

Lower Tropospheric and Surface Temperature Variations, 1958-2002

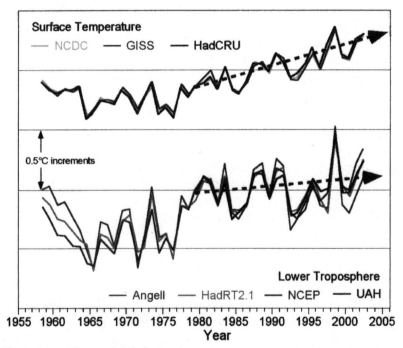

Figure 4.3.   History of global annual temperatures measured at the surface (upper curves) and measured in the lower atmosphere (lower curves) from 1958–2002 from a variety of data sets (HadCRU=Hadley Center Climate Research Unit surface measurements; NCDC=National Climatic Data Center surface measurements; GISS= Goddard Institute for Space Sciences surface measurements; HadRT2.1=version 2.1 of the Hadley Center Radiosonde Temperatures; UAH=University of Alabama-Huntsville Satellite-derived measurements; Angell=Jim Angell's radiosonde measure-ments; NCEP=National Center for Environmental Prediction radiosonde data set).

1989 La Niña), and cooling following large volcanic eruptions that inject sunlight-reflecting aerosols in the stratosphere (e.g., Mount Pinatubo, 1991–1992). Volcanic and ENSO events account for about 60 percent of the monthly to seasonal variability in the time series of tropospheric anomalies (Christy and McNider 1994; Michaels and Knappenberger 2000; Santer et al. 2001). In the stratosphere, volcanic aerosols cause substantial warming as the aerosols absorb energy from solar and upwelling infrared radiation. The overall negative trend in the lower stratosphere is consistent with the depletion of ozone (Christy and Drouilhet 1994; Folland et al. 2001). Little more will be said about the stratosphere as the process there, dominated by radiation, is fairly well known. It is the action of significant turbulence and the changing phases of water vapor that create immense complexity in the troposphere that lead to uncertainties there.

Numerous studies have been published in which these various data sets have been compared including IPCC and NRC reports. In terms of dependence, these data sets fall into two main groupings: (1) radiosonde based—NOAA Angell, NOAA LKS, NCEP, RIHMI, and HadRT, and (2) microwave based—UAH and RSS. But the methods by which each of these data sets is constructed differ to the extent that the largest disparities appear between time series that were derived from the same raw data rather than from independent data.

Greatest confidence regarding radiosonde-based data occurs in the lower tropospheric data sets, where information below 300 hPa is generally consistent, subject to fewer missing reports and minimally affected by instrumental changes. At higher levels, many factors work to enhance inhomogeneities, for example impacts between instruments of different manufacturers or sampling errors due to lost data. Those factors include solar radiation, with the associated unique correction tables, functionality in the extremely cold environment, balloon bursting, and loss of signal due to high velocity winds. Parker et al. found examples of shifts of 3 K in the stratosphere due to changes in instrumentation alone. Christy et al. (2003a) show the effect at Chuuk in which the LS temperature shifted downward by 1.34 K when the instrument was changed from VIZ-B to Vaisala (tropospheric shifts of that magnitude were not found). It is often difficult to know whether the raw measurements contributed to the shifts or whether the combination of new raw measurements and new correction tables supplied by the manufacturers were responsible. The uncertainty about the precise cause of the spurious shifts makes creation of homogeneous time series of radiosonde data based on a priori information especially challenging (Parker et al. 1997; Lanzante et al. 2003a, b).

One issue that creates uncertainty in global average calculations is the geographical sampling of the radiosonde stations used in each of the compilations. The number of stations respectively for NOAA-Angell, NOAA-LKS, and HadRT is 63, 87 and 400+, respectively. Hurrell et al. (2000) noted that calculating latitude-band anomalies with whatever stations were present in the particular latitude band was a preferable first step from which the global mean averages would be constructed. Though having fewer stations, the tropical bands will be better represented by this technique. Other averaging methods simply take the average of the available grids, which then skews the result to NH continental areas. Even so, the sparsity of the two NOAA networks is a concern regarding spatial sampling of the global average.

Lanzante et al. (2003b) performed a nonparametric comparison between the UAH and NOAA-LKS time series at the grid level, finding that the median trend differences for LT, MT, and LS were quite small, +0.00, −0.05, and −0.05°C decade^{-1} respectively (LKS minus UAH). These values indicate much closer agreement than found in trends of the global averages in table 4.2, perhaps due to the lack of spatial representativeness in LKS, and/or the influence on the overall average of a few particular stations with substantial differences. Indeed, of the twenty-one tropical comparisons, the UAH MT trend was more positive than LKS in fourteen of the cases. However, the trends for two LKS-corrected India stations were more positive compared with UAH MT by +0.7°C decade^{-1}, which would have a large impact on the average difference trend made up largely of values in the few hundredths of degree per decade. Because of difficulty with homogenizing India stations, the Hadley Centre excluded them from their HadRT2.X products (Folland et al. 2001).

The comparisons in table 4.2 indicate the microwave-based temperature trends are generally more positive than those of radiosonde-based products. The discrepancy increases as the layer's altitude increases, indicating that spurious stratospheric cooling related to instrument changes in the radiosondes is a likely cause. Thus, reliability of trends decreases as the layers of interest increases with the radiosondes. On the other hand, adjustments to the microwave-based data sets become more straightforward (e.g., less surface emissivity issues) as the altitude increases.

Santer et al. (2003) employed a curious methodology to assess the differences between UAH and RSS midtropospheric trends. Using an ensemble of climate model simulations with the known forcing of the past several decades, press reports concluded that RSS data were more likely to be accurate than UAH. Thus, in that scheme, a climate model was employed to test observed data. Santer et al. neglected to report the many other tropospheric data sets and studies that support the UAH long-term trends. They even cited one of the papers (Lanzante et al. 2003), yet failed to mention what the paper actually reported—there was very high consistency between the LKS trends and those of UAH. As a result, it is not useful to consider that methodology as addressing the issue of precision in long-term trends. For example, given all of the supporting evidence for slow tropospheric warming, a person could just as justifiably conclude that the climate model simulations in Santer et al. are likely erroneous in describing the changes in the climate system since 1978 (Christy et al. 2003b).

In all cases, the global and tropical trends of the lower troposphere since 1979 were more negative than those of the surface data sets, especially in the tropics. Since 1958, the surface and tropospheric trends are not significantly different, with the troposphere perhaps slightly more positive than the surface. As figure 4.3 demonstrates, the temperature shift in 1976–1977, also notable in the North Pacific atmospheric circulation, is the source of the tropospheric trend differences between the 1958–2002 and 1979–2002 periods (Folland et al. 2001).

Christy et al. (2001) demonstrated that sea-surface temperatures (SSTs) since 1979, measured anywhere from 1 m to 20 m depth and used in all "surface temperature" data sets, reveal greater warming than that of the air immediately above the ocean surface, measured on ship decks and buoys. Thus this evidence shows the rate of warming in the tropics is apparently greatest in the water, and decreases within the atmosphere.

Surface and Tropospheric Trend Differences in the Tropics

The issue of disparate trends between the surface and the troposphere in the tropics has attracted significant attention in the scientific community due to the inability of climate model simulations to reproduce this result (e.g., NRC 2000a; NAS 2001). In the following discussion, the focus will be on what research may be brought to bear on this trend issue; it is not intended to be an exhaustive study on the physical processes involved.

CLIMATE MODEL CONSIDERATIONS

Over the longer period of 1958 to 2002, tropical trend differences between surface and the LT were small. Keep in mind that one of the main conclusions from IPCC 2001 was that the surface warming of the second half of the past century (i.e., since 1950) was mostly due to human-induced changes in the atmospheric concentrations of greenhouse gases. But the surface temperature trend was zero from 1950 to 1979, indicating that all the global surface temperature increase of the second half of the twentieth century occurred in its last twenty-two years. That is precisely the period in which satellites began providing bulk temperature measurements, allowing a direct comparison with surface temperatures for the period in which human-induced change is thought to be most pronounced.

The simple way in which heat is vertically transported in climate models is based on the difference in the relative temperature between the planetary boundary layer and that of the troposphere above. As the surface warms relative to the troposphere in these convective parameterizations, the model atmosphere detects that instability and responds roughly with heating along the moist adiabatic lapse rate (about $-5°C$ km^{-1}), which is less steep than the typical environmental lapse rate (about $-6.5°C$ km^{-1}). That phenomenon produces, in the model atmosphere, what is known as the negative lapse rate feedback, where heat energy is transferred upward and deposited in the mid- and upper troposphere so that the temperature at these upper levels actually warms more rapidly than the surface in those grid cells so affected (NRC 2000a, 22; IPCC 2001, 715.) Other dynamic, diffusional, and radiational controls of the model then come into play to relax the horizontal and vertical temperature gradients that are created within those gridded columns. Even with those smearing factors, the mid- and upper troposphere warms more than the surface in the model tropically averaged atmosphere. Typically, the rate at which a model tropical troposphere warms is a factor 1.1 to 1.3 times that of the surface (e.g., Chase et al. 2004).

The observational evidence since 1979 appears to contradict the model simulations (Gaffen et al. 2000b; Brown et al. 2000). Assuming reasonable atmospheric data since 1958, the forty-five-year tropospheric and surface trends are not statistically different though uncertainties are larger in the pre-1979 upper-air data due to poor geographic representation where NH land stations dominated the network (table 4.2).[5] As a consequence, we focus on the post-1978 period. In the tropics, the convective parameterizations exert major controls of the model atmospheres and thus warm the troposphere more rapidly than the surface. But observations in the tropics since 1978 reveal a relative negative trend between the troposphere and surface over the past twenty-four years (about $0.15°C$ decade^{-1}). That is precisely the period of surface warming used to support the major conclusion of IPCC 2001 and a period where model results for the bulk of the atmosphere have little consistency with observations.

Chase et al. (2003) examined the probability of finding a significant difference in trends between the surface and troposphere over multiple periods of twenty-two years in length using four climate model long-term simulations. Experiments were studied for both unforced and forced (i.e., greenhouse-gas enhancement) integrations. *Their conclusions include the result that at no time, in any model realization, forced or unforced, did any model simulate the currently observed situation of a large and highly significant surface warming accompanied by little or no warming aloft.* Given the serious uncertainties in the factors that control convection and radiation, and considering the approximations employed in model simulations, I suggest, consistent with Chase et al., that these climate models fail to relinquish heat energy, which is apparently being incorrectly retained in their atmospheres. And it is therefore likely that long-term integrations of these models might not produce reliable estimates of atmospheric temperature trends.

Though the thrust of the National Academy of Sciences (NAS 2001) report promoted evidence for human induced climate change, its conclusion about this surface/troposphere disparity was less firm:

> The finding that the surface and troposphere temperature trends have been as different as observed over intervals as long as a decade or two is difficult to reconcile with our current understanding of the processes that control the vertical distribution of temperature in the atmosphere.

"Our current understanding" is a phrase that means the "results of climate model simulations." What the NAS statement implies, then, is that the inability to reproduce the behavior of so important a quantity as the bulk atmospheric temperature and the energy content it represents, is an inconsistency that should not be discounted when trying to predict how the climate system may unfold.

EXPLANATIONS BASED ON NATURAL FLUCTUATIONS

One possible explanation for the trend disparity relates to the representativeness of the past twenty-five years against the background of natural climate variability. Perhaps natural climate variations (i.e., unforced) could account for this disparity. (By implication then, it would follow that climate models in general do not contain proper natural variations.) The especially strong volcanoes of 1982 and 1991 coupled with significant warm ENSO events of 1982–1983 and 1997–1998 have considerable impact on the trend metric because the surface and troposphere exhibit differing magnitudes of thermal response. In general the troposphere responds with greater amplitude, while the surface, constrained by the large thermal inertia of the sea surface temperatures, displays a more muted impact.

For instance, Christy and McNider (1994) determined that Mount Pinatubo's eruption in 1991 induced a more negative trend in the 1979–1993 tropospheric time series than otherwise would be the case. The troposphere cooled by at least 0.7°C, while the surface temperatures fell by less than 0.4°C, so the cooling effect of the Mount Pinatubo aerosols was greater in the troposphere. Recently two studies report that even though the time series has lengthened, providing less potential for a single interannual event to impact the trend, nonetheless, these events continue to have noticeable influence. The significantly warm ENSO of 1997–1998 caused the 1979–1998 tropospheric trend to become more positive by about 0.05°C decade^{-1} compared with the trend ending in the year immediately prior (Santer et al. 1999; Michaels and Knappenberger 2000). When the influence of volcanoes and ENSOs was statistically removed from the surface and tropospheric time series, both studies found that the surface trend still exceeded that of the troposphere, though Michaels and Knappenberger demonstrated a greater remaining disparity than did Santer et al. Thus, the large interannual variations due to such events as volcanoes and El Niños do not explain the disparity.

Other studies investigated the influence of modes of repeating patterns of global atmospheric variability that create differential anomaly responses in the atmosphere versus the surface. Thompson et al. (2000) found differential impacts in the NH temperature structure of circulations associated with the high latitudes, but these were not evident globally or in annual averages. Hegerl and Wallace (2002) examined both ENSO and the Pacific Decadal Oscillation (another large-scale atmospheric pattern described by Frauenfeld in chapter 7 of this volume) but found little impact on the aforementioned trend disparity of these modes. Thus, an explanation of the trend differences appears not to rest on an hypothesis from the circulation argument either. Looking at all of the preceding results, it is clear that an explanation of the trend disparity has yet to arise from the modeling arena or from the natural variations of the system.

THERMAL REPOSITORIES: KEEPING TRACK OF ENERGY

How are these observations of trend disparity to be explained? Another way to view this issue is to keep track of the units of energy as they accumulate and are transferred between the various components in the global earth atmosphere system. If energy is being trapped by the enhanced greenhouse effect, it must be apparent somewhere. For example, with the bulk atmospheric temperatures available from microwave-based sources, we may estimate changes in energy stored in the atmospheric portion. To deal with the energy of the climate system in manageable numbers, given that the earth's surface area is 5.1×10^{14} m^2, I shall normalize the earth to a single area of one m^2 and utilize the approximation of 10^5 kg of atmosphere per m^2.

The key quantity needed to determine the change in energy content of the atmosphere is the change in mass-weighted temperature. (In this paradigm, the value of the surface temperature is relatively unimportant because it does not represent a meaningful reservoir of energy.) Using the LT, MT, and LS products of UAH, we calculate a twenty-four-year mean atmospheric temperature change of slightly less than zero (-0.026 K) as a consequence of significant stratospheric cooling versus modest tropospheric warming. Therefore, in the past quarter century, the number of joules of energy in the entire atmosphere has actually declined.[6]

The IPCC uses a convention in which the enhanced greenhouse effect is defined as the forcing at the top of the troposphere, not the full atmosphere. Again, using the LT and MT products, the estimated $+0.08°$C decade^{-1} full tropospheric trend translates to a gain of dry static atmospheric energy (i.e., internal and potential) of about 1.93×10^7 j m^{-2} since 1979. (Possible gains in kinetic and water vapor energy are neglected here.) In terms of mean forcing at the top of the troposphere (about 85 percent of the atmosphere by mass), that would account for the net absorption of a constant input of 0.021 Wm^{-2}.[7]

Levitus et al. (2001 and updates) estimate the ocean energy content has risen at a rate consistent with an increased absorption of slightly less than 0.2 Wm^{-2} since 1979, ranging from -0.6 to 1.2 Wm^{-2} in five-year averages. Levitus et al. also estimated the heat content change in other components (glacier melt, Arctic sea ice reductions, etc.) that suggest a rate of increase in the earth system heat content of these smaller reservoirs of the same order of magnitude as the troposphere or less. Thus the energy gain of the climate system since 1979 is dominated by the added heat content— joules gained—found in the upper ocean. One may estimate that all detectible changes in the climate system amount to no more than a net constant absorption of 0.25 Wm^{-2} since 1979 from these studies.

In keeping with the notion of counting joules of energy in the earth system, the IPCC 2001 estimates with high confidence that the current annual net increase in carbon dioxide forcing alone is about 1.5 Wm^{-2} at the top of the troposphere (about 10 km to 15 km altitude). No one takes issue with the fact that carbon dioxide concentrations have and continue to increase in the atmosphere. Other changes in radiative forcing, known with less confidence, are also positive—such as CH4, tropospheric O3, black carbon aerosols, and solar intensity. Still others are thought to exert negative forcing: stratospheric O3, sulfate, other aerosols, and changes in land use. The estimated annual total of these nonnatural forcings is about 1.2 Wm^{-2} since 1979 (table 6.13 in Ramaswamy et al. 2001.)

Given the observational estimate of about 0.25 Wm^{-2} as the rate of extra heat being retained within the earth system since 1979, we would expect the heat imbalance in models to reflect this 0.25 Wm^{-2} net absorption. What is problematic here though is that a model may generate an imbalance of 0.25 Wm^{-2}, which is miniscule compared with the background average of 235 Wm^{-2} for the wrong reason. In fact, some models show that the current balance is at least 0.8 Wm^{-2}. (Admittedly, the magnitude of the individual components of the estimated forcing changes is known with low confidence, so the summed values are somewhat speculative.)

This analysis suggests that about 0.25 Wm^{-2} has been retained in the earth system over and above the amount expected from a system in balance since 1979. If a highly accurate energy sensor were monitoring the earth from space, it would detect a decrease in outgoing energy due to the earth's greenhouse effect (all other things being equal) because those extra joules were being trapped in various components of the climate system as noted above rather than escaping to space. Over the latter part of the twentieth century, the detector would notice a tiny decline of outgoing energy relative to the background of 235 Wm^{-2} (if the incoming energy from the sun were absorbed at a constant rate). The typical scenario for the future, in which greenhouse gas concentrations continue to increase beyond the present day, requires that the radiational forcing will increase beyond these values, meaning more and more joules of energy will be trapped in the system somewhere. These joules would be expected to manifest themselves in higher ocean and atmospheric temperatures, diminishing glaciers and ice caps, and more evaporated water, among other things.

Evidence suggests that the amount of thermal radiation lost to space from the tropics, representing about one third of the earth's area, has actually increased since 1979 by about 4 Wm^{-2} (Hartmann 2002; Chen et al. 2002; Wielicki et al. 2002). The amount of incoming solar absorption over the same period in that region increased by only 1 to 2 Wm^{-2}. In other words, the measurements from space suggest a net loss of energy from the earth since 1979 at the same time that the components of the earth system have apparently gained energy. Thus, a contradiction seems to appear in which heat content measurements of the earth system point to an increase in energy stored, and measurements from space indicating (at least in the tropics) an increasing loss of energy. (I would caution that all energy measurements are subject to errors of at least the same magnitude as these changes being discussed, especially measurements from space.)

The key unknown here is whether the earth is experiencing a long-term imbalance of energy and how much that imbalance is. The models insist that today, more energy is being absorbed than expelled because that is the straightforward result of the basic and rather simple physical laws within the models. Some data suggest the opposite (more energy being ejected than absorbed) but energy data from satellites are relatively uncertain. Even so, one could argue that the climate system has responded in such a way as to counteract the imbalance of the net enhanced greenhouse forcing so that rather than being 0.8 Wm^{-2} at the moment as determined by models, it could be nearer zero or it could be close to the rate of the past 24 years (0.25). In other words, the actual climate system may have a way to reach a balance much more quickly than models suggest or have an imbalance due to other causes. The former could happen with only a very small change in certain parts of the system.

If the signs of the satellite observations of energy are correct, one possible explanation for the increased loss of thermal energy to space, in spite of increases in greenhouse gas forcing, is that over the past few decades changes in the distribution of cloudiness and/or upper tropospheric water vapor have opened up additional venting opportunities for thermal radiation to escape to space more readily rather than building up in the atmosphere. Lindzen et al. (2001) offered an explanation consistent with these observations indicating as SSTs increase, high cloudiness decreases, allowing more heat to escape. Their analysis has been challenged by others (e.g., Lin et al. 2002) and exchanges in the literature continue (Chou et al. 2002; Chambers et al. 2002). In any case, whatever reason may be proposed, it appears that the atmosphere has developed a means to shed heat at an increased rate proportional to the enhancement of the greenhouse effect. This finding is consistent with the lack of warming observed in the tropical troposphere since 1978 (table 4.2) and suggests the processes which lead to this heat loss are likely associated with the tropical atmosphere. Climate models have to date been unable to replicate this phenomenon, as Wielicki so aptly observes (see below).

The issue here is whether the main reservoirs of energy have been accurately identified and quantified, and how much energy will be transferred between them as time goes on. Given that the bulk of the enhanced energy storage has occurred in the ocean, one may suggest that atmospheric warming in the distant future is already foreordained as these joules will at some point be transferred to the air (e.g., Pielke 2003). In other words, joules of energy are being stored where they can be readily absorbed (e.g., the ocean), but this will create imbalances between this reservoir and others, such as the atmosphere, so that eventually those joules will move to reduce the imbalance. The response of the secondary reservoirs will become as evident as, for example, a warmer troposphere, melting ice caps, and so on.

Another view, based on empirical examination of forcing and response time constants, suggests that the atmosphere (and climate system in general) responds more quickly to enhanced forcing, and thus the impact of the enhanced energy storage is already essentially evident in the atmospheric system (Lindzen and Giannitsis 2002). In other words, imbalances created as one reservoir absorbs more energy than another are more quickly diffused to the adjacent reservoirs. In this view, the climate system's response is presently measurable within the time frame of the past few decades, providing the information necessary to calculate the impact of the enhanced greenhouse effect. Applying this notion using present observations suggests a relatively modest atmospheric warming of about 1°C over the next century. New results from proxy records suggest this is well within the atmospheric range of temperatures observed during the past two millennium (Soon and Baliunas 2003) and the past three interglacial periods (e.g., Watanabe et al. 2003).

It is likely that climate models do not properly store and transfer energy among the many climate system reservoirs, including the loss of energy to space. On the time frame of decades for example, Wielicki et al. note the lack of model replication of the observed level of interdecadal variability of energy escaping to space:

> We conclude that the [observed] large decadal variability of the LW [earth's heat loss] and SW [earth's heat gain from sun] radiative fluxes appear to be caused by changes in both the annual average and seasonal tropical cloudiness. In general, these changes are not well predicted by current climate models. Indeed, the current assessments of global climate change have found clouds to be one of the weakest components of climate models. This leads to a threefold uncertainty in the predictions of the possible global warming over the next century.

By attempting to track the "extra" energy absorbed into the earth-atmosphere system it appears that there is less being absorbed than anticipated from climate model projections. This is especially true in the atmosphere where processes may have evolved to allow much of the imbalance of the enhanced greenhouse forcing (the main factor in the imbalance calculated by models in 1979) to be mitigated. If this situation continues, one would expect a linear increase in temperature on the order of 0.1°C decade^{-1} in the bulk atmosphere. Since there have been serious questions raised about the magnitude of the future concentrations of anthropogenically produced greenhouse gases (i.e., the magnitude of future forcing) as well as the climate responses to that forcing, this issue of long term projections continues to be wrapped in considerable uncertainty.

NOTES

1. Due to the rapid horizontal mixing of the upper atmosphere, the geographical spacing of measurements is much greater than is required for the surface with its many incongruities.

2. MSU is the Microwave Sounding Unit and AMSU is the Advanced MSU.

3. During the writing of this chapter, two other microwave-based data sets have appeared which claim greater tropospheric warming than UAH (Vinnikov and Grody 2003; Fu et al. 2004). Neither of these have been assessed by independent evaluation and clearly depart from the results of those independent sources shown in table 4.2. Fundamental problems in both of these new data sets have been discussed in various open venues as official responses take many months.

4. IPCC 2001 (Folland et al. 2001) noted that surface and tropospheric trends were virtually identical over high-latitude continents such as North America and Europe.

5. Brown et al. (2000) and Gaffen et al. (2000b) indicate the tropospheric trend may have been more positive than that of the surface in the pre-1979 data. Lindzen et al. (2003), among many, note that the relative upward atmospheric temperature trend is an artifact of the sudden tropospheric warming in the climate shift of 1976–1977, and thus was not a gradual feature of the climate system.

6. A joule (j) is a unit of energy. About 1,000 joules added to 1 kg of air would increase its temperature by about 1°C. Again, it takes about 2.5 million joules to evaporate 1 kg of liquid water at 0°C. Our calculations deal with dry static energy, while the most accurate value would include the effect of moisture changes. However, moisture data sets for this period have large uncertainties.

7. A watt (W) is a rate of energy change, being one joule per second. There are 7.574 billion seconds in twenty-four years.

REFERENCES

Angell, J. K. 1988. Variations and trends in tropospheric and stratospheric global temperatures, 1958–87. *J Climate* 1:1296–1313.

Angell, J. K. 2000. Difference in radiosonde temperature trend for the period 1979–1998 of MSU data and the period 1959–1998 twice as long. *Geophys Res Lett* 27:2177–80.

Barnett, T. P. 1986. Detection of changes in the global tropospheric temperature field induced by greenhouse gases. *J Geophys Res* 91:6659–67.

Basist, A. N., and M. Chelliah. 1997. Comparison of tropospheric temperatures derived from the NCEP/NCAR reanalysis, NCEP operational analysis and the Microwave Sounding Unit. *Bull Amer Meteor Soc* 78:1431–47.

Brown, S. J., D. E. Parker, C. K. Folland, and I. Macadam. 2000. Decadal variability in the lower-tropospheric lapse rate. *Geophys Res Lett* 27:997–1000.

Chambers, L., B. Lin, B. Wielicki, Y. Hu, and K.-M. Xu. 2002. Reply. *J Climate* 15:2716–17.

Chase, T. N., R. A. Pielke Sr., B. Herman, and X. Zeng. 2004. Likelihood of rapidly increasing surface temperatures unaccompanied by strong warming in the free troposphere. *Clim Res* 25:185–90.

Chen, J., B. E. Carlson, and A. D. Del Genio. 2002. Evidence for strengthening of the tropical general circulation in the 1990s. *Science* 295:838–41.

Chou, M.-D., R. S. Lindzen, and A. Y. Hou. 2002. Comments on "The Iris hypothesis: A negative or positive cloud feedback?" *J Climate* 15:2713–15.

Christy, J. R., and S. J. Drouilhet. 1994. Variability in daily, zonal mean lower-stratospheric temperatures. *J Climate* 7:106–20.

Christy, J. R., and R. T. McNider. 1994. Satellite greenhouse signal. *Nature* 367:325.

Christy, J. R., R. W. Spencer, and R. T. McNider. 1995. Reducing noise in the MSU daily lower tropospheric global temperature data set. *J Climate* 8:888–96.

Christy, J. R., R. W. Spencer, and E. S. Lobl. 1998. Analysis of the merging procedure for the MSU daily temperature time series. *J Climate* 11:2016–41.

Christy, J. R., R. W. Spencer, and W. D. Braswell. 2000. MSU Tropospheric temperatures: Data set construction and radiosonde comparisons. *J Atmos Oceanic Tech* 17:1153–70.

Christy, J. R., D. E. Parker, S. J. Brown, I. Macadam, M. Stendel, and W. B. Norris. 2001. Differential trends in tropical sea surface and atmospheric temperatures. *Geophys Res Lett* 28:183–86.

Christy, J. R., R. W. Spencer, W. B. Norris, W. D. Braswell, and D. E. Parker. 2003a. Error estimates of Version 5.0 of MSU/AMSU bulk atmospheric temperatures. *J Atmos Oceanic Tech* 20:613–29.

Christy, J. R., R. W. Spencer, and W. D. Braswell. 2003b. Reliability of satellite data sets. *Science* 301:1046–47.

Christy, J. R., and W. B. Norris. 2004. What may we conclude about tropospheric temperature trends? *Geophys Res Lett* 31(6): L0621.

Eskridge, R. E., O. A. Alduchov, I. V. Chernykh, Z. Panmao, A. C. Polansky, and S. R. Doty. 1995. A Comprehensive Aerological Reference Data Set (CARDS): Rough and systematic errors. *Bull Amer Meteor Soc* 76:1759–75.

Folland, C. K., T. R. Karl, J. R. Christy, R. A. Clarke, G. V. Gruza, J. Jouzel, M. E. Mann, J. Oerlemans, M. J. Salinger, and S.-W. Wang. 2001. Observed climate variability and change. In *Climate change 2001: The scientific basis*. Contribution of Working Group I to the Third Assessment Report of the Intergovernmental Panel on Climate Change. Edited by J. T. Houghton, Y. Ding, D. J. Griggs, M. Noguer, P. J. van der Linden, X. Dai, K. Maskell, and C. A. Johnson. Cambridge: Cambridge University Press.

Free, M., I. Durre, E. Aguilar, D. Seidel, T. C. Peterson, R. E. Eskridge, J. K. Luers, D. Parker, M. Gordon, J. Lanzante, S. Klein, J. R. Christy, S. Schroeder, B. Soden, L. M. McMillin, and E. Weatherhead. 2002. Creating climate reference data sets: CARDS workshop on adjusting radiosonde temperature data for climate monitoring: Meeting summary. *Bull Am Met Soc* 83:891–99.

Fu, Q., C. M. Johanson, S. G. Warren, and D. J. Seidel. 2004. Contribution of stratospheric cooling to satellite-inferred tropospheric temperature trends. *Nature* 429:55–58.

Gaffen, D. J. 1993. Historical changes in radiosonde instruments and practices, WMO/TD-No. 541. *Instruments and observing methods report No. 50,* World Meteorological Organization, Geneva.

Gaffen, D. J. 1994. Temporal inhomogeneities in radiosonde temperature records. *J Geophys Res* 99:3667–76.

Gaffen, D. J. 1996. A digitized metadata set of global upper-air station histories. NOAA Technical Memorandum ERL ARL-211.

Gaffen, D. J., M. Sargent, R. E. Habermann, and J. R. Lanzante. 2000a. Sensitivity of tropospheric and stratospheric temperature trends to radiosonde data quality. *J Climate* 13:1776–96.

Gaffen, D. J., B. D. Santer, J. S. Boyle, J. R. Christy, N. E. Graham, and R. J. Ross. 2000b. Multi-decadal changes in the vertical temperature structure of the tropical troposphere. *Science* 287:1239–41.

Hartmann, D. L. 2002. Tropical surprises. *Science* 295:811–812.

Hegerl, G. C., and J. M. Wallace. 2002. *J Climate* 15:2412–28.

Hurrell, J., S. J. Brown, K. E. Trenberth, and J. R. Christy. 2000. Comparison of tropospheric temperatures from radiosondes and satellites: 1979–1998. *Bull Amer Met Soc* 81:2165–77.

IPCC (Intergovernmental Panel on Climate Change). 1996. *Climate Change 1995: The Science of Climate Change.* Edited by J. T. Houghton, F. G. Meira Filho, B. A. Callander, N. Harris, A. Kattenberg, and K. Maskell. Cambridge: Cambridge University Press.

IPCC (Intergovernmental Panel on Climate Change). 2001. *Climate Change 2001: The Scientific Basis.* J. T. Houghton and D. Yihui (co-chairs). Cambridge: Cambridge University Press.

Kalnay, E., M. Kanamitsu, R. Kistler, W. Collins, D. Deaven, I. Gandin, M. Iredell, et al. 1996. The NCEP/NCAR 40-year reanalysis project. *Bull Am Met Soc* 77:437–71.

Lanzante, J. R., S. A. Klein, and D. J. Seidel, 2003a. Temporal homogenization of monthly radiosonde temperature data. Part I: Methodology. *J Climate* 16:224–40.

Lanzante, J. R., S. A. Klein, and D. J. Seidel. 2003b. Temporal homogenization of monthly radiosonde temperature data. Part II: Trends, sensitivities and MSU comparison. *J Climate* 16:241–62.

Levitus, S., J. I. Antonov, J. Wang, T. L. Delworth, K. W. Dixon, A. J. Broccoli. 2001. Anthropogenic warming of the Earth's climate system. *Science* 292:267–70.

Lin, B., B. A. Wielicki, L. H. Chambers, Y. Hu, and K.-M Xu. 2002. The iris hypothesis: A negative or positive cloud feedback? *J Climate* 15:3–7.

Lindzen, R. S., M.-D. Chou, and A. Y. Hou. 2001. Does the earth have an adaptive iris? *Bull Amer Met Soc* 82:417–32.

Lindzen, R. S., and C. Giannitsis. 2002. Reconciling observations of global temperature change. *Geophys Res Lett* 29(12):1583–86.

Luers, J. K., and R. E. Eskridge. 1998. Use of radiosonde temperature data in climate studies. *J Climate* 11:1002–19.

Mears, C. A., M. C. Schabel, and F. J. Wentz. 2003. A Reanalysis of the MSU Channel 2 Tropospheric Temperature Record. *J Climate* 16:3650–64.

Michaels, P. J., and P. C. Knappenberger. 2000. Natural Signals in the MSU lower tropospheric temperature. *Geophys Res Lett* 27:2905–8.

NAS. 2001. *Climate change science: An analysis of key questions.* Washington, D.C.: National Academy Press.

NRC. 1999. *Adequacy of climate observing systems.* Washington D.C.: National Academy Press.

NRC. 2000a. *Reconciling observations of global temperature change.* Washington D.C.: National Academy Press.

NRC. 2000b. *Issues in the integration of research and operational satellite systems for climate research II. Implementation.* Washington D.C.: National Academy Press.

Oort, A., and H. Liu. 1993. Upper-air temperature trends over the globe. *J Climate* 6:292–307.

Parker, D. E., and D. I. Cox. 1995. Towards a consistent global climatological raw insonde data-base. *Int Journ Clim* 15:473–96.

Parker, D. E., M. Gordon, D. P. N. Cullum, D. M. H. Sexton, C. K. Folland, and N. Rayner. 1997. A new global gridded radiosonde temperature data base and recent temperature trends. *Geophys Res Lett* 24:1499–1502.

Pielke, R. A., Sr. 2003. Heat storage within the earth system. *Bull Amer Meteor Soc* 84:331–35.

Pielke, R. A., Sr., J. Eastman, T. N. Chase, J. Knaff, and T. G. F. Kittle. 1998. 1973–1996 Trends in depth-averaged tropospheric temperature. *J Geophys Res* 16:927–33.

Ramaswamy, V., O. Bucher, J. Haigh, D. Hauglustaine, J. Haywood, G. Myhre, T. Nakajima, et al. 2001. Radiative forcing of climate change. In *Climate change 2001: The scientific basis.* Contribution of Working Group I to the Third Assessment Report of the Intergovernmental Panel on Climate Change. Edited by

J. T. Houghton, Y. Ding, D. J. Griggs, M. Noguer, P. J. van der Linden, X. Dai, K. Maskell, and C. A. Johnson. Cambridge: Cambridge University Press.

Santer, B. D., J. J. Hnilo, T. M. L. Wigley, J. S. Boyle, C. Doutriaux, M. Fiorino, D. E. Parker, and K. E. Taylor. 1999. Uncertainties in observationally based estimates of temperature change in the free atmosphere. *J Geophys Res* 104(D6): 6305–33.

Santer, B. D., T. M. L. Wigley, C. Doutriaux, J. S. Boyle, J. E. Hansen, P. D. Jones, G. A. Meehl, et al. 2001. Accounting for the effects of volcanoes and ENSO in comparisons of modeled and observed temperature trends. *J Geophys Res* 106:28033–60.

Santer, B. D., K. E. Taylor, J. S. Boyle, C. Doutriaux, T. M. L. Wigley, G. A. Meehl, C. Ammann, et al. 2003. Influence of satellite data uncertainties on the detection of externally forced climate change. *Science Express*, May 1.

Seidel, D. J., J. K. Angell, J. Christy, M. Free, S. A. Klein, J. R. Lanzante, C. Mears, et al. 2004. Uncertainty in signals of large-scale climate variations in radiosonde and satellite upper-air temperature data sets. *J Climate*. Submitted.

Soon, W., and S. Baliunas. 2003. Proxy climatic and environmental changes of the past 1,000 years. *Clim Res* 23:89–110.

Spencer, R. W., and J. R. Christy. 1990. Precise monitoring of global temperature trends from satellites. *Science* 247:1558–62.

Spencer, R. W., and J. R. Christy. 1992a. Precision and radiosonde validation of satellite gridpoint temperature anomalies. Part I: MSU channel 2. *J Climate* 5:847–57.

Spencer, R. W., and J. R. Christy. 1992b. Precision and radiosonde validation of satellite gridpoint temperature anomalies. Part II: A tropospheric retrieval and trends during 1979–90. *J Climate* 5:858–66.

Spencer, R. W., and J. R. Christy. 1993. Precision lower stratospheric temperature monitoring with the MSU: Validation and results 1979–91. *J Climate* 6: 1194–1204.

Spencer, R. W., J. R. Christy, and N. C. Grody. 1990. Global atmospheric temperature monitoring with satellite microwave measurements: Methods and results 1979–84. *J Climate* 3:1111–28.

Stendel, M., J. R. Christy, and L. Bengtsson. 2000. Assessing levels of uncertainty in recent temperature time series. *Climate Dynamics* 16(8):587–601.

Sterin, A. M. 1999. An analysis of linear trends in the free atmosphere temperature series for 1958–1997. *Meteorologiai Gidrologia* 5:52–68.

Thompson, D. W. J., J. M. Wallace, and G. C. Hegerl. 2000. Annual modes in the extratropical circulation Part II: Trends. *J Climate* 13:1018–36.

Thorne, P. W., D. E. Parker, S. F. B. Tett, P. D. Jones, M. McCarthy, H. Coleman, P. Brohan, and J. R. Knight. 2005. Revisiting radiosonde upper-air temperature from 1958–2002. *J Geophys Res* (submitted).

Trenberth, K. E., and J. G. Olson. 1991. Representativeness of a 63-station network for depicting climate changes. In *Greenhouse-gas-induced climatic change: A critical appraisal of simulations and observations.* Edited by M. E. Schlesinger. San Diego: Elsevier Science.

Trenberth, K. E., D. P. Stepaniak, J. W. Hurrell, and M. Fiorino. 2001. Quality of reanalyses in the tropics. *J Climate* 14:1499–1510.

Vinnikov, K., and N. Grody. 2003. Global warming trend of mean tropospheric temperature observed by satellites. *Sciencexpress,* September 11, 2003.

Watanabe, O., J. Jouzel, S. Johnsen, F. Parrenin, H. Shoji, and N. Yoshida. 2003. Homogeneous climate variability across East Antarctica over the past three glacial cycles. *Nature* 422:509–12.

Wentz, F. J., and M. Schabel. 1998. Effects of satellite orbital decay on MSU lower tropospheric temperature trends. *Nature* 394:361–64.

Wielicki, B. A., T. Wong, R. P. Allan, A. Slingo, J. T. Kiehl, B. J. Soden, C. T. Gordon, et al. 2002. Evidence for large decadal variability in the tropical mean radiative energy budget. *Science* 295:841–44.

5

SEVERE WEATHER, NATURAL DISASTERS, AND GLOBAL CHANGE

Randall S. Cerveny

An enormous disconnection exists between popular perception and scientific reality regarding the effects of global climate change on severe weather and natural disasters. With the occurrence of almost any major weather-related disaster anywhere in the modern world, global news web pages are likely to have a sidebar story as to whether that disaster has specific linkages to global change. For example, media reports on massive heat waves in Europe (summer 2003), a sustained outbreak of tornadoes in North America (spring 2003), even a bitter cold wave in the northeastern United States (winter 2003–2004) have all been cited as potential indicators of global climate change. With 24/7 news coverage (television, print media, electronic, etc.) of these events (a phenomenon that is remarkable in and of itself), it is not surprising that many people believe severe weather events have increased in frequency and in damage over recent decades and will continued to do so into the future.

The scientific world, however, is much less strident in producing such "predictions." This chapter examines the scientific evidence for changes in severe weather as a function of global climate change—in particular, severe storm events occurring in the United States. In its *Glossary of Meteorology*, the American Meteorological Society defines severe storms as specific events (e.g., tornadoes, hurricanes, and thunderstorms), as opposed to longer-term episodes (e.g., heat waves, droughts); it is that definition that pertains here.

IS SEVERE WEATHER ON THE RISE?

Is there any scientific basis for believing that severe weather is increasing because of anthropogenic climate change? Many sources cite a buildup of greenhouse gases for a pronounced increase in near surface air temperature measurements over the past fifty years. Others voice equally strong reservations about the accuracy of such measurements.

If we accept, for discussion purposes, the a priori assumption that greenhouse gases are having a direct and visible impact on global temperatures, then we can speculate on what scientific linkages can be made to severe weather. Two general types of severe weather would likely be affected by global warming: tropical cyclones (including tropical storms and hurricanes) and extratropical thunderstorms (including air mass thunderstorms and nor'easters). Both general types are strongly influenced by convective processes and atmospheric instability.

First, on a simple theoretical level, a general warming of the lower atmosphere would likely enhance the chances for convection. Convection is the basic process by which heat energy is transferred in the atmosphere. Hot air will rise, condense into clouds and, under certain circumstances, generate precipitation and potentially severe weather. The larger the thermal difference between the surface and upper atmosphere, the greater the convection and the more unstable the atmosphere will be. Consequently, given that higher lower-atmosphere temperatures enhance convection and therefore precipitation (and that all other factors are held constant), theoretically, general global warming should produce warmer surfaces and that would, through convection, lead to greater precipitation.

In particular, global warming has been associated with warmer nights (Easterling et al. 1997). Given the premise of global warming, that diurnal finding, along with the observed tendency for thunderstorms to occur at night throughout much of the interior of the country (Wallace 1975), then, again theoretically, convection should be enhanced at night over much of the United States.

But one problem with such a simplistic argument is that it does indeed assume "all other factors are held constant." In other words, the temperature convection argument assumes one-to-one relationships; that is, that warmer temperature only impacts convection, and convection only impacts thunderstorm intensity and frequency. A simplistic argument of this type ignores the very real possibility of negative feedback mechanisms. A feedback mechanism is where a change in the climate system leads to either an amplification (positive feedback) or weakening (negative feedback) of the original change. For example, while hotter surface temperatures can lead to more convection, convection will produce clouds. Sustained or long-lasting cloud cover would, however, eventually reflect more sunlight and lead to a cooling of the surface, thereby weakening convection. Many other potential negative and positive feedbacks exist in our atmosphere–indeed, some may still be undiscovered.

A similar temperature convection–storm intensity theoretical situation exists with tropical cyclones, such as tropical storms and hurricanes. At its most fundamental level, a tropical cyclone is a massive heat engine whose power source is the release of heat that has been locked up in water through the process of condensation. Consequently, many scientists have speculated that warmer sea-surface temperatures (SSTs) created in a greenhouse world would lead to warmer surface air. Warmer surface air can hold more moisture, so a warmer world should produce larger and more frequent hurricanes, right?

A major series of research papers is fundamentally responsible for that speculation. Kerry Emanuel theorized in a series of articles published in the mid- to late 1980s (1986, 1987, 1988) that elevated sea-surface temperatures could increase the potential destructive power of tropical storms by 40 percent to 50 percent, thereby creating supertropical storms that he termed "hypercanes." Subsequent research using sophisticated computer experiments has tended in part to support Emanuel's conclusion. Indeed, many numerical modeling papers have demonstrated that a warmer world with higher sea-surface temperatures and elevated atmospheric moisture levels could increase the frequency, intensity, or duration of future tropical cyclones or alter the mean location of the storms (e.g., O'Brien et al. 1992; Ryan et al. 1992; Haarsma et al. 1993; Evans et al. 1994; Whitney and Hobgood 1997; Knutson et al. 1998; Royer et al. 1998; Druyan et al. 1999; Knutson and Tuleya 1999; Walsh and Katzfey 2000; Walsh and Ryan 2000; Knutson et al. 2001; Sugi et al. 2002; Tsutsui 2002). For example, Druyan et al. (1999) and Sugi et al. (2002) have used general circulation models to compute potential hurricane activity for a doubled CO_2 world; their models produced, specifically in the Atlantic Ocean, a marked increase in storm activity. In another modeling study (Knutson et al. 2001), elevated CO_2 increased maximum hurricane wind speeds by up to 10 percent and produced significant rainfall increases compared to present-day conditions.

But not all numerical model simulations suggest increased hurricane activity given a continued buildup of greenhouse gases. For example, a numerical study using the National Oceanic and Atmospheric Administration Geophysical Fluid Dynamics Laboratory hurricane model found that hurricanes were less intense when created under global warming scenarios with increased CO_2 (Shen et al. 2000). Even those studies showing hurricane increases have surprise findings. For example, the model simulations by Sugi et al. (2002) mentioned above as showing an increase in Atlantic tropical cyclone activity also identified an overall significant reduction in the frequency of tropical cyclones in response to the greenhouse gas-induced global warming. The Sugi et al. study determined that the most significant decrease is indicated over the North Pacific Ocean, the ocean basin associated with greatest tropical cyclone activity. A subsequent modeling study by Sugi and Yoshimura (2004) has found that, if sea-surface temperature is held constant, a doubled CO_2 atmosphere reduces radiative cooling in the lower troposphere, thereby leading to an overall reduction in tropical precipitation; but, if potential SST increases are taken into account, the increase in atmospheric temperature and water vapor would lead to increases in both radiative cooling and overall tropical precipitation. Consequently, although not all numerical experiments lead to the "prediction" of increased hurricane activity given the continued buildup of greenhouse gases (e.g., Shen et al. 2000), a theoretical basis for "hypercanes" can be identified in the literature.

OBSERVED EVIDENCE

If there is a theoretical basis for suggesting greater storm activity, both tropical and extratropical storms, does the available evidence confirm a connection between greenhouse warming and storm activity? As noted earlier, many sources cite a buildup of greenhouse gases for a pronounced increase in near surface air temperature measurements over the past fifty years. Consequently, an examination of observed scientific measurements of storm activity over the past fifty years should aid in determining the actual impact of enhanced atmospheric greenhouse gases of severe storm characteristics.

Rainfall

First, over the past one hundred years, rainfall derived from thunderstorms has indeed increased over most of the United States (Changnon 2001). Other research, employing a variety of statistical and classification analyses, has indicated that overall precipitation levels are increasing in the country, with much of that increase associated with occurrence of heavy and extreme daily precipitation events (Karl et al. 1995; Karl and Knight 1998; Groisman et al. 2001). Specifically, researchers have determined that the heaviest rainfall events have shown the greatest increases over the past fifty years. Those events, together with the overall increases in rainfall, have significantly affected river flow, including a higher probability of floods (Groisman et al. 2001).

Thunderstorms

But is that demonstrated increase in United States rainfall mirroring overall changes in severe storm activity? Let's first examine overall thunderstorm activity. When thunderstorm occurrence (the observation of actually hearing thunder at a given weather station) is charted over the past hundred years for eighty-six high-quality weather stations, researchers identified three basic trends. They found that approximately one third of the stations had an upward trend in thunderstorm activity, one third had a downward trend, while the other third had no trend whatsoever (Changnon and Changnon 2001). When the stations were categorized by region, the researchers discovered that only the western part of the United States showed any increase in thunderstorm activity through the past century. Indeed, when all eighty-six stations were averaged together, the resulting national average showed a slight decline over time. It is important to note that no increases were seen in the Great Plains and Midwest regions, where some of the most severe thunderstorms occur.

Potentially, critics might suggest that extrapolating those findings to the more basic question of "increasing severe weather" is difficult given the rather limited and somewhat subjective definition of "thunderstorm" as an occurrence of audible thunder. More comprehensive examination of severe thunderstorms should involve the specific elements of severe thunderstorms. These include the phenomena of hail and tornadoes.

Hail

Two general changes in hail incidences over the past century can be identified from sixty-six high-quality weather stations across the United States. The first is an overall increase in hail activity from early in the past century (1916) to around 1955, when the highest values were achieved. The second change, however, is an overall decline in hail activity from 1955 to 1995 (Changnon and Changnon 2001). The net distribution in hail activity over the past century is that of a bell-shaped curve with a peak in the midpart of the century.

Tornadoes

A second measure of severe thunderstorm activity is tornado activity. Of course, a major problem with recording tornadoes is the increase in population for tornado-prone regions over the past fifty years. For example, while total tornado reports have increased by a factor of ten in the past fifty years (e.g., Grazulis 1993; Brooks and Doswell 2001), most, if not all, of that increase is due to better observation and reporting procedures (such as the use of Doppler radar), particularly of weaker tornadoes, as well as a greater observing population (Brooks and Doswell 2001). Eliminate consideration of weaker tornadoes (because of the technological improvements in detecting them throughout the past century), and limit consideration to only the most severe tornadoes (using the standard ranking system developed by Dr. Theodore Fujita; F3 tornadoes and greater), and no observable trend in severe tornado occurrence is evident over the past fifty years (Balling and Cerveny 2003).

Nor'easters and Blizzards

Of critical importance to the heavily metropolitan east coastal area are the severe coastal winter storms that are commonly called nor'easters because of their prevailing wind direction. Several recent attempts to classify nor'easters, using a variety of indicators to define winter storms, concur that no discernible trends exist in the occurrence of such storms over the past fifty years. For example, when winter storms are defined using storm surge (coastal flooding) records, the resulting analyses of storm frequency indicate considerable interdecadal variability but no visible long-term trend in the number and intensity of moderate or severe coastal storms during this century (Zhang et al. 2000). Similarly, a study of severe east coast winter storms found no significant trends in winter storm frequency over a forty-six-year period beginning in 1951 (Hirsch et al. 2001). In fact, the one trend that was marginally statistically significant was a slight increase in the average minimum surface pressure of nor'easters (Hirsch et al. 2001). If that trend is indeed occurring, a rise in nor'easter surface pressures indicates an overall *weakening* of such storms.

A second type of winter storm, characterized by heavy snow and ice, high winds and blizzard conditions, and dangerous wind-chill events, can occur over wide expanses of the continental regions of the United States. Analyses of the number of midcontinental winter storms from 1979 to 1993 identified significant interannual and monthly variabilities but did not specifically examine long-term trends (Dery and Yau 1999). An analysis of damaging or injurious winter events including seasonal fluctuations, relative frequencies of different hazards (e.g., heavy snow vs. freezing precipitation), duration variations, and size (areal coverage) distributions also determined interannual and monthly variability (Branick 1997).

The geographic extent, variety of associated phenomena (e.g., ice, wind, etc.), and within-event variability accent the difficulties in studying trends in midcontinental blizzards.

Hurricanes

Scientific debate on changes in tropical storms and hurricanes has raged between experiments involving numerical modeling and those involving analysis of historical records. Again, many numerical modeling simulations have concluded that a greenhouse-warmed world would increase the frequency, intensity, or duration of tropical cyclones. In contrast, with the exception of the last six years of the twentieth century, hurricane and tropical storm records over the past century have exhibited no upward trend or even declining levels of tropical storm frequency and intensity in the specific types of tropical cyclones examined by those modeling simulations.

For example, Wilson (1999) examined intense hurricanes in the Atlantic basin from 1950 to 1998 and found a decreasing trend and fewer intense hurricanes during the warmer periods. Easterling et al. (2000) reported a general decline in landfalling Atlantic hurricanes in the latter part of the twentieth century. In the case of the United States, however, Elsner et al. (2000) warned that the Arctic warming and the relaxation of the North Atlantic Oscillation may reverse those downward trends and bring an increase in North American hurricane activity in the near future.

But those last six years of the last century (1995 to 2000) are troubling and have raised alarms. In those six years, the North Atlantic Ocean experienced the highest level of hurricane activity ever recorded in modern records. Scientists identified a 2.5-fold increase in major hurricanes (category 3 or higher on the Saffir-Simpson hurricane intensity scale), and a fivefold increase in hurricanes affecting the Caribbean, compared with the previous quarter century (Elsner et al. 2000; Goldenberg et al. 2001). Is this a sign of climate change?

The consensus of hurricane researchers is no. The greater activity is thought to have resulted from simultaneous increases in North Atlantic sea-surface temperatures and decreases in vertical wind shear that are associated with multidecadal natural circulation shifts and not necessary from long-term climate change (Goldenberg et al. 2001).

Given the lack of evidence for changes in frequency of hurricanes, another aspect of hurricane activity—one that hurricane policy makers and insurance companies have discussed—is whether greenhouse warming could impact how fast a hurricane intensifies. For example, in 1996, Hurricane Lili increased its wind from 47 ms^{-1} (~105 mph) to 60 ms^{-1} (~134 mph) within 15 hours. Does the historical record reveal any significant indication of more rapidly intensifying hurricanes in recent times?

Analysis of tropical cyclone records from the Caribbean, Gulf of Mexico, and tropical sector of the western North Atlantic over the past fifty years disclose no significant trends in the start date, ending date, or duration of the storm season (defined in various ways) and no trends in the average geographic position of the storms (Balling and Cerveny 2003). Furthermore, various measures of hurricane season timing and storm locations were not related to regional sea surface temperature or the northern hemispheric or global temperatures.

So, what conclusions can be made with regard to hurricanes and climate change? Raghavan and Rajesh (2003) recently stated in the prestigious science journal *Bulletin of the American Meteorological Society*: "Contrary to the common perception that tropical cyclones are on the increase, due perhaps to global warming, studies all over the world show that, although there are decadal variations, there is no definite long-term trend in the frequency or intensity of tropical cyclones over the period of about a century for which data are available."

"NO SYSTEMATIC CHANGE . . ."

In summary, with the exception of heavy rain events, the research consensus on individual severe weather phenomena is that severe weather is not increasing. But is that message reaching the entire scientific community? For the answer to that question, we must look to the premier scientific voice addressing climate change in the world today, the United Nations Intergovernmental Panel on Climate Change (IPCC). Does their published opinion correspond to scientific data presented above? Indeed, it does.

In a recent report by that group, the IPCC scientists concluded "no systematic changes in the frequency of tornadoes, thunder days, or hail events are evident in the limited areas analyzed" (Houghton et al. 2001, 5). When the IPCC science group evaluated the evidence on climate change in tropical cyclones and hurricanes, it warned that hurricane modeling studies involving climate change are compromised by a variety of uncertainties and scaling problems, and therefore assigned only a "medium likelihood" that changes will occur in the frequency, intensity, or location of tropical cyclones (Houghton et al. 2001, 528). The IPCC scientists concluded that "changes globally in tropical and extratropical storm intensity and frequency are dominated by interdecadal and multidecadal variations, with no significant trends evident over the twentieth century" (Houghton et al. 2001, 104).

WHY DON'T WE HEAR THAT MESSAGE?

Scientific consensus exists. The facts of the matter are clear. But why don't we hear them? There are several reasons: Nonspecialists charged with disseminating scientific information that may arguably be difficult to comprehend; media eager to fill airtime or column inches by stirring controversy or conforming to a political or social agenda; and the general public's own sense of personal history that past eras "just weren't this bad."

Media Filters

One of the fundamental causes for the public perception of increasing natural disasters is that such scientific information must be filtered through an increasing network of global communications and assessments by nonspecialists. Competing and 24/7 continuous media coverage of news, including natural disasters, often leads to discussion of environmental issues in forums that may not correspond to the scientific facts but rather to political or social agendas. That sort of reporting—which reaches well beyond recounting facts and into the realm of often uninformed speculation—unfortunately often leads to misinterpretation of the results. As respected climatologist Ann Henderson-Sellers noted in a critical review of the media coverage of a climate change panel's findings on greenhouse-induced tropical cyclone change (1998, 421), "the two days of media coverage essentially rewrote the committee findings."

As an additional example, in March 2004, a leading global insurance agency (Swiss Re) produced a report entitled *Natural Catastrophes and Man-Made Disasters in 2003*. Authors of that report stated that storms and extreme weather events would increase in the future, citing the simplistic argument that a one-to-one relationship exists between temperature distributions and severe weather distributions. They suggested that if the distribution of maximum daily temperatures were to change, then (apparently assuming no changes in severe weather distribution), the number of severe weather events must increase. Such an argument disregards the existence of climate feedback relationships that can exist between surface temperature increases and severe weather as well as the observed evidence cited in this chapter. Nevertheless, the Reuters news agency stated that "insurance giant

Swiss Re warned . . . that the costs of natural disasters, aggravated by global warming, threaten to spiral out of control, forcing the human race into a catastrophe of its own making. . . . Scientists expect global warming to trigger increasingly frequent and violent storms, heat waves, flooding, tornadoes, and cyclones while other areas slip into cold or drought." Reuters included no opposing information to that report, nor did the news agency list the names of the scientists involved in preparing it.

Increased media coverage of natural disasters has had one markedly positive, though little publicized, benefit: the total number of tornado deaths in the United States has significantly decreased over the past few decades. While some might immediately suggest that is a positive long-term climatic change, it is much more likely that that downward trend in tornado deaths is fundamentally related to greater public awareness of the danger of tornadoes and dissemination of warnings (Grazulis 1993).

A Sense of History

Another problem with the public perception of increasing natural disasters is that people tend to have a relatively short-term memory with regard to these sorts of events. We tend to remember the most recent disasters and, unfortunately, forget older—and frequently deadlier—ones. For instance, many people regard the F5 tornado that struck Oklahoma City in May 1999 as one of the worst tornadoes ever. Not to minimize a deadly weather event, but they have already forgotten the superoutbreak of tornadoes in 1974, in which 148 tornadoes (three of which were F5 category) struck the Midwest over a terrifying twenty-four-hour period. Similarly, Hurricane Andrew in 1992 was undoubtedly a terrible disaster; twenty-three people died in the United States as a result of it. When the stronger Hurricane Camille struck the Gulf Coast in 1969, however, 143 people along the coast from Alabama into Louisiana were killed by the storm. Finally, the heat wave of 1995 in Chicago killed 599 people, a number many people in the United States regarded as unprecedented, not realizing that during a nine-day period in July 1939, 679 people died of heat-related problems in Michigan (Blumenstock 1959).

YET WE CONTINUE TO SEE MORE NATURAL DISASTERS

The media tends to dramatize changes in natural disaster and therefore plays a role in the public perception of more and larger natural disasters. But can that view of natural disasters be reconciled with the scientific facts about se-vere storm trends? Fundamentally, the question must be addressed: "Are we indeed experiencing more natural disasters today than we have in the past?"

Interestingly, and perhaps surprisingly, given all the negative links peo-ple make to climate change, the scientific answer to the question, Are we having more natural disasters? is unquestionably yes. However—and this is a very important qualification—the primary reason for the increase in disas-ters is *not* increased frequency or intensity of the natural phenomenon (tor-nado, hurricane, etc.) at the heart of a disaster. In other words, in general we are *not* experiencing more tornadoes, hurricanes, blizzards, or other nat-ural phenomena. The key lies in how we define the term "natural disaster."

There are two parts of any natural disaster: the event itself and the im-pact on people. Consider: A tree that falls in a vast forest isn't a natural dis-aster, but a tree that falls on a building full of people could be. Disasters, by definition, must happen to *people*. So if the core events (tornadoes, hurri-canes, etc.) are not increasing, why are we having more natural disasters?

Global population increases—and an increasing desire to live in areas prone to disasters (e.g., coastal areas susceptible to hurricanes)—are the fun-damental causes for increased natural disasters. Quite simply, many more people are living in disaster-prone regions. Consequently, the chances of a natural disaster—a natural event that leads to massive loss of human life and property—are higher because the population of the planet continues to in-crease, particularly in disaster-prone areas. For example, when hundreds of thousands of people live within a few feet of sea level in the delta area of the Bay of Bengal—especially when they have poor communication channels—the potential for a disaster is magnified regardless of the relative intensity of the cyclone that initiated it.

This point is emphasized in a study that specifically addressed trends in economic and human health impacts from weather and climate extremes (Kunkel et al. 1999). In terms of severe weather and natural disasters, the report's authors note that "increasing losses are primarily due to increasing vulnerability arising from a variety of societal changes, including a growing population in higher risk coastal areas and large cities, more property sub-ject to damage, and lifestyle and demographic changes subjecting lives and property to greater exposure" (Kunkel et al. 1999, 1077). They continue by noting that although fatalities and economic damages have increased due to severe storms, "when changes in population, inflation, and wealth are con-sidered, there is instead a downward trend."

Similarly, noted climatologist Stanley Changnon and colleagues de-veloped a set of national indices for weather-related losses that were all adjusted to 1997 dollars. They concluded, "Trends were upward for certain key variables between 1950 and 1997, including the incidence and losses associated with winter storms, flood losses, crop losses, and incidence of heavy rains. Trends were downward for other weather-driven loss variables, including hurricane losses, energy costs, thunderstorm losses, wind storm losses, and hail losses" (Changnon et al. 2001, 1). In like fashion, Pielke and Landsea (1998) addressed damage from hurricanes and found that the upward trend in losses in recent decades disappeared when they controlled for inflation and population growth in coastal areas.

THE BOTTOM LINE

Fundamentally, the weight of evidence regarding the impact of increasing atmospheric carbon dioxide and global warming on severe weather suggests that (1) climate change impact, at the present time, is small, if present at all, and is very difficult to adequately identify; (2) media sources and political interests have tended to exaggerate climate change's impact on severe weather in ways that correspond with the general public's faulty recollection of weather events and natural disasters in recent memory; and (3) population growth and redistribution, rather than actual climate change, is largely responsible for the observed increases in the frequency and severity of most "natural disasters."

REFERENCES

Balling, R. C., Jr., and R. S. Cerveny. 2003. Compilation and discussion of trends in severe storms in the United States: Popular perception v. climate reality. *Natural Hazards* 29(2):103–12.

Blumenstock, D. I. 1959. *The Ocean of Air*. New Brunswick, N.J.: Rutgers University Press.

Branick, M. L. 1997. A climatology of significant winter-type weather events in the contiguous United States, 1982–94. *Weather and Forecasting* 12:193–207.

Brooks, H. E., and C. A. Doswell. 2001. Normalized damage from major tornadoes in the United States: 1890–1999. *Weather and Forecasting* 16:168–76.

Changnon, S. A. 2001. Thunderstorm rainfall in the conterminous United States. *Bull Amer Met Soc* 82:1925–40.

Changnon, S. A., and D. Changnon. 2001. Long-term fluctuations in thunderstorm activity in the United States. *Clim Change* 50:489–503.

Dery, S. J., and M. K. Yau. 1999. A climatology of adverse winter-type weather events. *J Geophys Res* 104:16657–672.

Druyan, L. M., P. Lonergan, and T. Eichler. 1999. A GCM investigation of global warming impacts relevant to tropical cyclone genesis. *Int Journal Clim* 19:607–17.

Easterling, D. R., B. Horton, P. D. Jones, T. C. Peterson, T. R. Karl, D. E. Parker, M. J. Salinger, et al. 1997. Maximum and minimum temperature trends for the globe. *Science* 277:364–67.

Easterling, D. R., J. L. Evans, P. Y. Groisman, T. R. Karl, K. E. Kunkel, and P. Ambenje. 2000. Observed variability and trends in extreme climate events: A brief review. *Bull Amer Met Soc* 81:417–25.

Elsner, J. B., T. Jagger, and X.-F. Niu. 2000. Changes in the rates of North Atlantic major hurricane activity during the 20th century. *Geo Research Lett* 27: 1743–46.

Emanuel, K. A. 1986. An air–sea interaction theory for tropical cyclones. Part I: Steady-state maintenance. *J Atmos Sci* 43:585–604.

Emanuel, K. A. 1987. The dependence of hurricane intensity on climate. *Nature* 326:483–85.

Emanuel, K. A. 1988. The maximum intensity of hurricanes. *J Atmos Sci* 45:1143–55.

Evans, J. L., B. F. Ryan, and J. L. McGregor. 1994. A numerical exploration of the sensitivity of tropical cyclone rainfall intensity to sea surface temperatures. *J Climate* 7:616–23.

Goldenberg, S. B., C. W. Landsea, A. M. Mestas-Numez, and W. M. Gray. 2001. The recent increase in Atlantic hurricane activity: Causes and implications. *Science* 293:474–79.

Grazulis, T. P. 1993. *Significant tornadoes 1680–1991: A chronology and analysis of events*. St. Johnsbury, Vt.: Environmental Films.

Groisman, P. Y., R. W. Knight, and T. R. Karl. 2001. Heavy precipitation and high streamflow in the contiguous United States: Trends in the twentieth century. *Bull Amer Met Soc* 82:219–46.

Haarsma, R. J., J. F. B. Mitchell, and C. A. Senior. 1993. Tropical disturbances in a GCM. *Clim Dynamics* 8:247–57.

Henderson-Sellers, A. 1998. Climate whispers: Media communication about climate change. *Clim Change* 40:421–56.

Hirsch, M. E., A. T. DeGaetano, and S. J. Colucci. 2001. An East Coast winter storm climatology. *J Climate* 14:882–99.

Houghton, J. T., Y. Ding, D. J. Griggs, N. Noguer, P. J. van der Linden, X. Dai, and K. Maskell (Eds.). 2001. *Climate change 2001: The scientific basis*. Cambridge: Cambridge University Press.

Karl, T. R., and R. W. Knight. 1998. Secular trends of precipitation amount, frequency, and intensity in the United States. *Bull Amer Met Soc* 79:231–41.

Karl, T. R., R. W. Knight, D. R. Easterling, and R. G. Quayle. 1995. Trends in U.S. climate during the twentieth century. *Consequences* 1:3–12.

Knutson, T. R., and R. E. Tuleya. 1999. Increased hurricane intensities with CO_2-induced warming as simulated using the GFDL hurricane prediction system. *Clim Dynamics* 15:503–19.

Knutson, T. R., R. E. Tuleya, W. Shen, and I. Ginis. 2001. Impact of CO_2-induced warming on hurricane intensities as simulated in a hurricane model with ocean coupling. *J Climate* 14:2458–68.

Knutson, T. R., R. E. Tuleya, and Y. Kurihara. 1998. Simulated increase of hurricane intensities in a CO_2-warmed climate. *Science* 279:1018–20.

Kunkel, K. E., et al. 1999. Temporal fluctuations in weather and climate extremes that cause economic and human health impacts: A review. *Bull Amer Met Soc* 80:1077–98.

O'Brien, S. T., B. P. Hayden, and H. H. Shugart. 1992. Global climatic change, hurricanes, and a tropical forest. *Clim Change* 22:175–90.

Pielke, R. A., Jr., and C. Landsea. 1998. Normalized hurricane damages in the United States: 1925–1995. *Weather and Forecasting* 13:351–61.

Raghavan, S., and S. Rajesh. 2003. Trends in tropical cyclone impact: A study in Andhra Pradesh, India. *Bull Amer Met Soc* 84:635–37.

Royer, J.-F., F. Chauvin, B. Timbal, P. Araspin, and D. Grimal. 1998. A GCM study of the impact of greenhouse gas increase on the frequency of occurrence of tropical cyclones. *Clim Change* 38:307–43.

Ryan, B. F., I. G. Watterson, and J. L. Evans. 1992. Tropical cyclone frequencies inferred from Gray's yearly genesis parameter: Validation of GCM tropical climates. *Geophys Res Lett* 19:1831–34.

Shen, W., I. Ginis, and R. E. Tuleya. 2000. A sensitivity study of the thermodynamic environment on GFDL model hurricane intensity: Implications for global warming. *J Climate* 13:109–21.

Sugi, M., and J. Yoshimura. 2004. A mechanism of tropical precipitation change due to CO2 increase. *J Climate* 17:238–84.

Sugi, M., A. Noda, and N. Sato. 2002. Influence of the global warming on tropical cyclone climatology: An experiment with the JMA global model. *J Met Soc Japan* 8:249–72.

Swiss Re Reinsurance Company. 2004. *Natural catastrophes and man-made disasters in 2003: Many fatalities, comparatively moderate insured losses. Report No. 1.* Zurich: Swiss Reinsurance Company.

Tsutsui, J. 2002. Implications of anthropogenic climate change for tropical cyclone activity: A case study with the NCAR CCM2. *J Met Soc Japan* 80:45–65.

Wallace, J. M. 1975. Diurnal variations in precipitation and thunderstorm frequency over the conterminous U.S. *Monthly Weather Review* 103:406–19.

Walsh, K. J. E., and B. F. Ryan. 2000. Tropical cyclone intensity increase near Australia as a result of climate change. *J Climate* 13:3029–36.

Walsh, K. J. E., and J. J. Katzfey, 2000. The impact of climate change on the poleward movement of tropical cyclone-like vortices in a regional climate model. *J Climate* 13:1116–32.

Whitney, L. D., and J. S. Hobgood. 1997. The relationship between sea surface temperatures and maximum intensities of tropical cyclones in the eastern North Pacific Ocean. *J Climate* 10:2921–30.

Wilson, R. M. 1999. Statistical aspects of major (intense) hurricanes in the Atlantic basin during the past 49 hurricane seasons (1950–1998): Implications for the current season. *Geophys Res Lett* 26:2957–60.

Zhang, K. Q., B. C. Douglas, and S. P. Leatherman. 2000. Twentieth-century storm activity along the U.S. East Coast. *J Climate* 13:1748–61.

6

PRECIPITATION AND THE "ENHANCED" HYDROLOGIC CYCLE

David R. Legates

Warmer temperatures will lead to a more vigorous hydrological cycle; this translates into prospects for more severe droughts and/or floods in some places and less severe droughts and/or floods in other places. Several models indicate an increase in precipitation intensity, suggesting a possibility for more extreme rainfall events.

—Intergovernmental Panel on Climate Change, 1996

The importance of the hydrologic cycle to life on earth cannot be overstated. If changes in that cycle occur, potential impacts could be greater—both in terms of human lives and economic losses—than those arising from just an increase in air temperature alone. The Intergovernmental Panel on Climate Change (IPCC), in its *Second Assessment Report* (SAR), heralded the possibility that rising global temperatures would result in an enhanced hydrologic cycle characterized by more extreme conditions. Taken from its Summary for Policymakers, the passage quoted above provides no forecast for whether drought and flood frequencies will change overall—it says only that both positive and negative changes are likely to occur in a variety of places. But it does argue that an increased intensity of rainfall is likely.

The *Third Assessment Report's* (TAR) Summary for Policymakers, however, goes further: "Global warming is likely to lead to greater extremes of drying and heavy rainfall and increase the risk of droughts and floods that occur . . . in many different regions" (IPCC 2001).

That summary goes on to suggest that observations indicate precipitation has become more intense over many areas in the middle and upper latitudes of the Northern Hemisphere, as has summer continental drying, leading to enhanced drought frequencies. That is likely to continue, the TAR states, with more frequent drought being "very likely" during the twenty-first century. Although no changes in tropical cyclone mean and peak precipitation intensities have been observed, the TAR suggests that such observations are likely to increase over some areas during the twenty-first century.

As water proceeds through a series of phase changes—solid, liquid, vapor—it absorbs and releases energy. Evaporation or sublimation of water (a transition to vapor form directly from ice), for example, stores energy that is subsequently released when it condenses. Thus, the movement of water vapor in the atmosphere is an important energy transport mechanism. The process of condensation produces clouds that can simultaneously cool the planet through reflection of solar radiation and warm it through the absorption of terrestrial radiation. Water vapor is the most important, but often overlooked, greenhouse gas.

Basic theory suggests that rising air temperatures might be accompanied by a slight increase in globally averaged precipitation. The moisture content of the air at saturation increases with increasing air temperatures (figure 6.1). It therefore stands to reason that there will be the potential for slightly more water vapor in a slightly warmer atmosphere. The IPCC TAR agrees, although biases and errors in the data make it difficult to access significance of potentially small trends. Regionally, however, precipitation is affected by the atmospheric circulation and the transport of moisture, among other factors, so it is difficult to argue directly that an increase in air temperature should directly lead to an increase in precipitation for any given area.

BIASES IN PRECIPITATION MEASUREMENT AND THEIR INFLUENCE ON PRECIPITATION TRENDS

One of the problems with precipitation measurements is that the traditional use of a can-type precipitation gauge provides a biased estimate of precipitation. Such biases most often result from obstructions to the wind flow by the gauge itself, wetting losses on the internal walls of the gauge, evaporation from the gauge, and limitations of automatic recording techniques (see Legates 1987; Legates and Willmott 1990). With the gauge itself obstructing the wind, the air is forced to flow up and across the gauge opening, or orifice, resulting in an increase in the wind speed due to the compression of

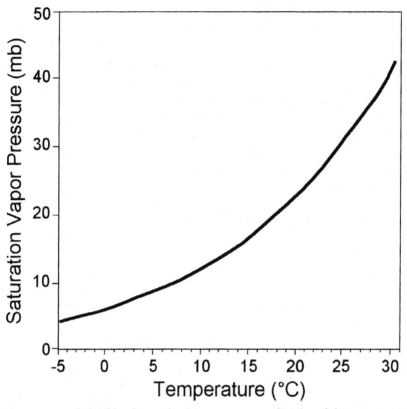

Figure 6.1. Relationship of saturation vapor pressure as a function of air temperature.

the air flow, slight updrafts across the gauge orifice, and a pressure difference between air at rest inside the orifice and air in motion across the gauge orifice. Those three factors result in a decrease in the gauge catch as wind speed increases, an effect that is more pronounced with snowfall than with rain. Legates and DeLiberty (1993) estimated that for much of the United States, the winter bias in precipitation measurement often exceeds 25 percent, while summer bias is less than 8 percent. Seasonal and interannual variations in both wind speed and air temperature (which change the proportion of precipitation falling as snow) affect the measurement bias, but so too do changes in instrumentation, local siting characteristics (e.g., obstructions such as trees and buildings), and station location. The magnitude of the bias is affected by these latter changes as they affect the wind speed across the gauge orifice, which, in turn, alters the magnitude of the gauge bias.

When changes occur suddenly, their effect may be easily identified as a discontinuity in the precipitation record. Often these changes are easily detected (Eishied et al. 1991; Groisman and Legates 1994, 1995), particularly if the change is dramatic, and are removed from the record, rather than being interpreted as a climate change signal. Subtle effects may go undetected, although when good station metadata (station history information) exist, their presence may still be noted. Changes in siting characteristics that occur slowly over long periods of time, however, are far more difficult to identify. Young trees may grow over several decades to become substantial windbreaks, providing a slow but significant change to the gauge environment that often is not documented in station metadata. Urbanization similarly can introduce a subtle but important change in the gauge environment that can erroneously be interpreted as a climate change signal.

Legates (1995) examined the effect of changing air temperature (affecting the proportion of snowfall in the total precipitation) and wind speed on the catch of a hypothetical gauge located in the Upper Midwest. He used a model of gauge measurement biases, a fairly modest 1°C per century increase in air temperature, and a 0.5144 ms^{-1} (1 knot) per century decrease in wind speed—assumed to have resulted possibly from the development of the urban heat island and increased obstructions in the urban environment. Legates demonstrated that the percentage increase in the gauge catch due solely to a reduction in the gauge measurement bias was about 6.4 percent per century. That means that even though it was assumed that precipitation, although variable, exhibited the same mean value for a century, the measured precipitation total, as recorded by the gauge, would have exhibited an increase of 6.4 percent per century simply as a result of the modest changes to air temperature and wind speed that were imposed.

Thus, subtle changes in the environmental conditions associated with a precipitation gauge can induce significant trends in the observed record. The nature of the precipitation gauge measurement bias, then, which is often erroneously considered to simply be a constant by many researchers, can in fact induce spurious trends into a precipitation time series. It therefore is quite difficult to ascertain how much of an observed trend is due to changes in the precipitation climate and how much of it arises from changes in the siting characteristics associated with the precipitation gauge.

Factors Affecting Flood and Drought Perception

One of the difficulties in addressing whether the hydrologic cycle is becoming more extreme lies both in the definition and perception of floods and droughts. A "flood condition" is almost always linked to stream flow above a prescribed threshold and not necessarily to increased precipitation.

Rainfall frequencies may increase, or the amount of rainfall per event may increase, but unless it results in an increase in river discharge, a "flood" in the traditional sense may not occur. Similarly, stream flow depends strongly on the antecedent moisture condition of the soil and detention areas—rain falling on dry soil or when lake levels are low may have little effect on river discharge, but that same amount of rain falling on saturated ground or when lakes are near capacity may produce substantial increases in stream flow. Moreover, stream flow in the early spring is greatly affected by snowmelt, which, in turn, depends on the rapidity of springtime warming and the winter accumulation of the snowpack. Thus, flood occurrences depend a great deal on the timing of precipitation and the form (rain or snow) in which the precipitation falls.

Changing land use conditions and engineering developments also affect the frequencies by which enhanced stream flows occur. With increased urbanization comes a change from vegetation to pavement and other more impervious surfaces that generate more overland runoff. Channelization efforts meant to direct streams and rivers, such as the construction of levees, dredging activities, and the reinforcement of stream banks, speed water flow and restrict water from entering the natural flood plain. Anthropogenic effects on rivers, therefore, often enhance the occurrence of flood conditions, masking the effect of potential changes in the earth's climate.

Droughts likewise are difficult to define. Most simply, a "meteorological drought" results when below-normal rainfall has occurred. Given the highly variable nature of precipitation, however, that definition strongly depends on the selected time period and the period over which "normal" rainfall has been defined. It is quite possible, for example, to have had a meteorological drought over the past six months, but have above-normal rainfall for the past year. The implication of that definition is that the drought ends only when the precipitation, summed from the beginning of the arbitrarily defined time period, equals the normal precipitation for the time period. In many cases, a flood can occur yet not end the meteorological drought. Thus, this definition of drought is seriously lacking.

An "agricultural drought" occurs when soil moisture conditions fall below normal levels to some degree. That definition is quite useful, as it has direct relationship to plant growth and crop yield, although for most agricultural areas, soil moisture is almost always replenished to field capacity during the dormancy of winter. For that reason, an agricultural drought rarely extends for longer than a single growing season. In more of a direct contrast with the definition of a flood, a "hydrological drought" occurs when river and lake levels (and sometimes well levels) fall below a specified threshold. Water for urban use almost always comes from surface or ground-water reserves, which makes that definition particularly useful as well. But with increased urbanization comes an increased demand for water use—industrial, residential, and agricultural—which leads to an increase in the occurrence of hydrologic drought conditions that also masks the effect of potential changes in the earth's climate.

Thus, our perception of drought and flood frequencies is greatly affected by anthropogenic influences that do not, in fact, affect the climate. With increased urbanization comes an increased demand for water, more impervious surfaces, and a channelization of streams and rivers that lead to increased frequencies of both hydrological droughts and floods. Even with an unchanging but variable climate, urbanization leads to an increase in these hydrological extremes. Thus, it is important to separate out the perception of an increase in the hydrological cycle with the true changes that have been observed with respect to precipitation and evapotranspiration.

Modeling Precipitation with a General Circulation Model

General circulation models (GCMs) attempt to describe the full three-dimensional structure of the earth's climate. They are used in a variety of applications, most notably the investigation of the possible role of various climate forcing mechanisms and the simulation of past and future climates. Although it appears that such models exhibit the potential to simulate changes to the real climate, several important issues must be considered. First, GCMs are limited by our incomplete understanding of the climate system and how the various atmospheric, land surface, oceanic, and ice components interact with one another. The climate system is extremely complex and we are only beginning to understand some of the effects, feedbacks, and interactions among its components. But even if a perfect understanding did exist, GCMs are further limited by our limited ability to transform this knowledge into a mathematical representation. While, for example, scientists may have a general idea of the complex interrelationships between the atmosphere and the oceans, converting that idea into a set of mathematical equations that can be solved analytically is much more difficult.

Second, GCMs are further limited by their own spatial and temporal resolutions. Processes in the real world operate on a variety of scales—from the molecular to a global spatial scale and from near instantaneous to geologic time scales. Finite restrictions on computing power and computational complexity reduce GCM simulations to coarse generalities with spatial resolutions no finer than at least one hundred kilometers. Consequently, many small-scale features—both in space and time—are not represented, even though they may exert a significant impact on the local, regional, or even global climate. Climate models consider land surfaces as being spatially uniform over large areas and flows of moisture and energy between the land surface and the atmosphere are averaged over these areas. But the extensive heterogeneity of the land surface and the effects that even small-scale changes can have make modeling land–surface interactions extremely challenging. At best, GCMs can only represent a gross thumbnail sketch of the real world. Regional assessments over areas encompassing many GCM grid cells are the finest spatial resolution that can be expected. It is inappropriate, and grossly misleading, to select results from a single grid cell and apply them locally. It cannot be over emphasized that GCM representations of the climate can be evaluated at a spatial resolution no finer than large regional areas, defined by a thousand square kilometers (at least several GCM grid cells) on a side. Even the use of "nested grid models" (models that take GCM output and resolve it to finer-scale resolutions) does not overcome this limitation, since results from the GCM simulation drive such models and the simulation receives no feedback of the results of such finer-scale models. Nor do most "statistical downscaling" techniques (methods by which GCM output is statistically adjusted to match observed means and variances), as such approaches tacitly assume that the local climate is merely a reflection of the larger scale pattern, save for a change in the statistical moments (e.g., means and variances) of the distribution.

A third and important limitation in GCMs is that, given our imperfect knowledge of the climate system as well as the computational complexities associated with modeling the climate system, GCMs simply cannot reproduce many important phenomena. Hurricanes and most other forms of severe weather (e.g., thunderstorms, tornadoes, and nor'easters) cannot be represented in a GCM owing to its coarse spatial resolution. Even weather fronts that are commonplace in midlatitudes are not simulated. Thus GCM simulated precipitation results not from known processes such as fronts and weather systems, but rather as a consequence only of atmospheric instability or convection. That leads to precipitation taking on a popcorn-like appearance, with disorganized patterns appearing continually.

More complex phenomena that result from interactions among the various components of the climate system may be badly represented, if they are represented at all. For example, El Ninõ and La Ninã phenomena, the Pacific Decadal Oscillation, PNA (Pacific–North American) teleconnection patterns, and other complex interrelationships are inadequately reproduced or often completely absent in climate model simulations. To be accurate, a climate model should exhibit those phenomena as a result of the specification of climate interactions and forcing mechanisms. Their absence indicates a fundamental flaw in either our understanding of the climate system, in our mathematical representation of the process, in the spatial and temporal limitations imposed by finite computational power, or any combination of the above.

A final limitation in climate modeling is that in the climate system, everything is interconnected. In short, anything you do wrong in a climate model will adversely affect the simulation of virtually every other variable and process. To cause air to condense and form precipitation requires moisture in the atmosphere and a mechanism to force it to condense (i.e., by forcing the air to rise over mountains, by surface heating, as a result of weather fronts, or by cyclonic rotation). Any errors in representing either the atmospheric moisture content or these precipitation-causing mechanisms will result in errors in the simulation of precipitation. For example, inaccuracies in the representation of topography will hinder an accurate simulation of regional-scale precipitation since mountains force air to rise and condense to produce orographic (mountain-induced) precipitation (e.g., the coastal mountain ranges of Washington and Oregon). Incorrect simulations of air temperature also will lead to inappropriate precipitation patterns since the ability of the atmosphere to store moisture is directly related to its temperature. If winds, air pressure, and atmospheric circulation are inadequately represented, then the simulation of precipitation will be adversely affected since the atmospheric flow of moisture that may condense into precipitation will be incorrect. Plant transpiration and soil evaporation also provide moisture for precipitation; therefore, errors in the simulation of soil moisture conditions and vegetation cover will adversely affect precipitation patterns. Simulation of clouds that alter the amount of solar energy reaching the ground will affect estimates of surface heating that, in turn, adversely affects the simulation of precipitation; particularly because precipitation-causing mechanisms in a GCM are limited to only atmospheric instability or convection. Even problems in specifying oceanic circulation or sea ice concentrations will affect weather patterns, which affect precipitation simulations. In sum, the simulation of precipitation is adversely affected by inaccuracies in the simulation of virtually every other climate variable and even specifications in land surface conditions.

Inaccuracies in precipitation simulations, in turn, adversely affect virtually every other climate variable. Condensation releases heat to the atmosphere and forms clouds, which reflect energy from the sun and trap heat from the earth's surface—both of which affect the simulation of air temperature. As a result, this affects the simulation of winds, atmospheric pressure, and atmospheric circulation. Since winds drive the circulation of the upper layers of the ocean, the simulation of ocean circulation also will be adversely affected. Sea ice formation in a climate model will be affected by inaccuracies in the air temperature simulations. As precipitation is the only source of soil moisture, inadequate simulations of precipitation will adversely affect soil moisture conditions and land surface hydrology. Vegetation also responds to precipitation availability, which means that the entire representation of the biosphere will be adversely affected. Clearly, the interrelationships among the various components that comprise the climate system make climate modeling difficult.

It is important to keep in mind that it is not just the long-term average and seasonal variations that are of interest in estimating the extent of climate change. Demonstrating that precipitation is highest over the tropical rainforests and lowest in the subtropical deserts is not enough for an adequate model. Climate change is likely to manifest itself in small regional fluctuations. Moreover, we also are interested in interannual (year-to-year) variability. Much of the character of the earth's climate is in how it varies over time. A GCM that simulates essentially the same conditions year after year—a condition that is inherent in most climate models—clearly is missing an important component of the earth's climate. Thus, the evaluation of climate change prognostications using GCMs must be made in light of the model's ability to represent the holistic nature of the climate and its variability.

An assessment of the efficacy of any climate model, therefore, must focus on the ability of the model to simulate the present climate conditions. If a model cannot simulate what we know to be true, then it is unlikely that model prognostications of climate change are believable. A word of caution, however. It is common practice to "tune" climate models so that they better resemble present conditions of air temperature. That is a widely accepted practice, because many parameters in GCMs cannot be specified directly, and their values must be determined through empirical trial and error. But if a GCM has been "tuned" to replicate current climate, can it be said to adequately simulate the processes that create that climate? A model may appear to provide a good simulation of air temperature, for example, when in fact the model may poorly simulate climate change mechanisms. In other words, a GCM may provide an adequate simulation of the present-day climate conditions but for the wrong reasons. Model efficacy in simulating present-day conditions, therefore, is not a guarantee that model-derived climate change scenarios will be reasonable.

With respect to model tuning and GCM limitations, consider the impact of inaccuracies in precipitation simulations on the simulation of air temperature, for example. The energy released by condensing moisture can be described by

$$\Delta P \, \rho_w \, L \tag{1}$$

where ΔP is the amount of moisture condensed (rainfall depth—volume of water per area of the earth's surface), ρ_w is the density of water, and L is the latent heat of vaporization. Changes in air temperature are governed by

$$\Delta T \, C_p \, p \, / \, g \tag{2}$$

where ΔT is the change in air temperature, C_p is the specific heat at constant pressure, p is the atmospheric pressure, and g is the gravitational acceleration. If the energy released by condensing moisture changes the temperature of the atmosphere, then Equation (1) is set equal to Equation (2). Solving for ΔT,

$$\Delta T = \Delta P \, (\rho_w \, L \, g \,) \, / \, (C_p \, p \,). \tag{3}$$

With appropriate choices of the various constants and integrating over the troposphere (lowest 80 percent of the mass of the atmosphere), it can be determined that a 1 mm error in simulating the precipitation rate equates to an error in the simulated air temperature of 0.39°C.[1] In English units, an error in precipitation of only 0.1 inch equates to an error of 1.77°F. This is important to remember when examining biases in GCM simulations of air temperature and precipitation.

The IPCC Claims for an Enhanced Hydrologic Cycle

Relying largely on GCM prognostications, the IPCC *Third Assessment Report* has made a number of claims, often contradictory, about changes in the hydrologic cycle as a result of anthropogenic increases in atmospheric trace gases. With respect to the hydrologic cycle, these include changes in total precipitation, precipitation intensity, floods and droughts, and storminess.

Changes in Precipitation Totals

It is very likely that precipitation has increased by 0.5 percent to 1 percent per decade in the twentieth century over most

mid- and high latitudes of the Northern Hemisphere continents. . . . It is likely that rainfall has increased by 0.2 percent
to 0.3 percent per decade over the tropical (10°N to 10°S)
land areas. Increases in the tropics are not evident over the past
few decades. . . . It is also likely that rainfall has decreased over
much of the Northern Hemisphere sub-tropical (10°N to
30°N) land areas during the twentieth century by about 0.3
percent per decade. . . . In contrast to the Northern Hemisphere, no comparable systematic changes have been detected
in broad latitudinal averages over the Southern Hemisphere.
There are insufficient data to establish trends in precipitation
over the oceans.

—IPCC 2001, 4

Based on global model simulations and for a wide range of
scenarios, global average water vapor concentration and precipitation are projected to increase during the twenty-first century. By the second half of the twenty-first century, it is likely
that precipitation will have increased over northern mid- to
high latitudes and Antarctica in winter. At low latitudes there
are both regional increases and decreases over land areas. Larger
year to year variations in precipitation are very likely over most
areas where an increase in mean precipitation is projected.

—IPCC 2001, 13

As precipitation and air temperature are the two most widely measured
meteorological variables, it stands to reason that long-term changes in precipitation totals would garner nearly as much attention as has long-term
trends in air temperature. The problem, though, is that precipitation is far
more variable, exhibiting considerable interannual and spatial fluctuations.
Coupled with the inherent bias in measuring precipitation and the near
complete lack of precipitation for the world's oceans, it is quite difficult to
divine long-term trends from the high degree of variability or "climate
noise" present. Nevertheless, several recent studies have examined trends in
precipitation totals for annual, seasonal, or monthly time scales.

For global land areas, New et al. (2001) conclude that terrestrial precipitation (excluding Antarctica) has increased by only about 9 mm over
the past century—a modest trend of less than 1 mm per decade (figure
6.2). The greatest observed trend is in the Southern Hemisphere, where an
increase of 22.4 mm has been observed, in contrast to the Northern Hemisphere, which exhibits an increase of only 3.2 mm. Even so, none of these

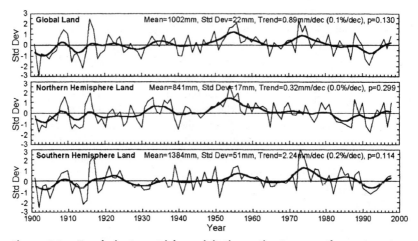

Figure 6.2. Trends in terrestrial precipitation estimates over the past century. Trends are shown for the entire globe (top), the Northern Hemisphere (middle), and the Southern Hemisphere (bottom). Note that the mean and standard deviation are different for each plot. None of these trends, however, is statistically significant (adapted from New et al. 2001).

trends is statistically significant, as the decadal variability is more than four times this amount. Statistically significant trends were observed poleward of 40N, however. New et al. further argue that all parts of the globe have experienced increased precipitation, with the exception of parts of southern Africa, Amazonia, western South America, and tropical North Africa. Of course, such analyses are spatially limited to less than a third of the earth's surface, owing to the underrepresentation of high latitudes and mountainous regions and the near complete absence of observations over the world's oceans. Since 1980, when spatially complete assessments of global precipitation became available, few regions show marked trends in precipitation (New et al. 2001).

In a widely cited study, Karl and Knight (1998) examined precipitation trends for the continental United States. They concluded that mean precipitation has increased by about 10 percent since 1910—a figure that is much greater than the results of New et al. (2001) for the same latitudinal band. No trend was observed in median precipitation, about which Karl and Knight conclude that most of the change has occurred in the heaviest precipitation events. Similarly, Mekis and Hogg (1999) found a 10 percent increase in precipitation in Canada over the past century, a finding echoed by Akinremy et al. (1999) for the Canadian prairies and by Zhang et al. (2000) for heavy snowfall in northern Canada.

With respect to model simulations of precipitation, Meehl et al. (2000, 430) state very succinctly that "increased precipitation intensity (albeit with certain regional variations) in a future climate with increased greenhouse gases was one of the earliest model results regarding precipitation extremes, and remains a consistent result with improved, more detailed models." Using the climate models featured in the United States National Assessment, however, Doherty and Mearns (2000) succinctly demonstrated that for North America, the difference between the model simulated precipitation and the observations is considerable (see figure 6.3, where results for the summer months are shown). Comparisons between the Legates-Willmott (Legates 1987; Legates and Willmott 1990) global precipitation climatology and the Hadley Centre (HadCM2) and Canadian Global Coupled (CGCM1) models (model runs chosen for use in the United States National Assessment), demonstrate that model simulations of present-day conditions can differ from the observations by as much as the average precipitation. For comparison, the model prognostications by 2030 are shown. Note that the model prognostications are dwarfed by the large errors in the model simulations. Such substantial errors make it exceedingly difficult to glean reliable climate change scenarios from these models. Thus, the IPCC (2001) conclusion that "based on global model simulations and for a wide range of scenarios, global average . . . precipitation [is] projected to increase during the twenty-first century" is tenuous, at best (see also Felzer and Heard 1999).

Changes in Precipitation Frequency and Intensity

> In the mid- and high latitudes of the Northern Hemisphere over the latter half of the twentieth century, it is likely that there has been a 2 percent to 4 percent increase in the frequency of heavy precipitation events.
>
> —IPCC 2001, 4

Analyzing precipitation frequencies and intensities is possibly more rewarding than examining long-term averages (Trenberth et al. 2003). The IPCC argues that an increase in precipitation totals will occur as a result of changes in the frequency and intensity of precipitation events. In the global terrestrial analysis of New et al. (2001), evidence is presented for increased daily precipitation resulting from an increase in the frequency of days with precipitation and an increased proportion of precipitation that results from the heaviest rainfall intensities.

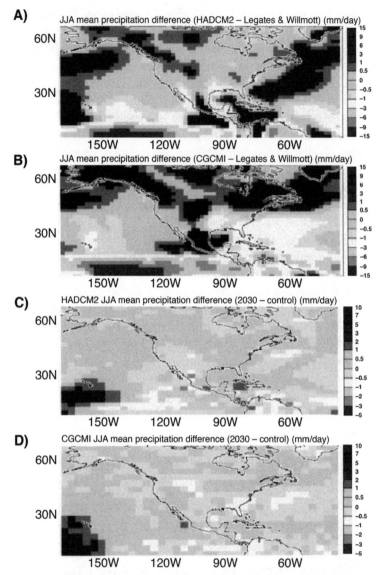

Figure 6.3.   Differences between model-simulated and observed (Legates and Willmott 1990) summer precipitation as simulated by the (a) Hadley Climate Model (HadCM2) and (b) the Canadian Global Coupled Model (CGCM1). Magnitudes of the projected changes by 2030 are shown for the two models (c and d), respectively, to illustrate the magnitude of the signal relative to the model uncertainties (from Doherty and Mearns 2000).

Much of the work in this area, however, focuses on precipitation in the United States. Karl and Knight (1998) concluded that the frequency of days with precipitation increased for all precipitation categories—an amazing increase of 6.3 days per century. Note that this finding is, in fact, contrary to their earlier argument that there was no change in median precipitation. Nevertheless, 53 percent of the total precipitation increase occurred with the heaviest precipitation category—daily precipitation exceeding 50 mm (2 inches). Nationally, however, moderate precipitation intensities decreased by 1 percent. But it was the increase in the number of days with precipitation that accounted for 87 percent of the total precipitation increase.

They concluded, "These data suggest that the precipitation regimes in the United States are changing disproportionately across the precipitation distribution. The proportion of total precipitation derived from extreme and heavy events is increasing relative to more moderate events" (231). Similar conclusions were found or echoed by Karl et al. (1996), Easterling et al. (2000), and Meehl et al. (2000). Groisman et al. (1999) too found similar results, although their use of an artificial frequency distribution (i.e., the Gamma distribution) rather than observed class frequencies adversely affects their results. Dai (1999) further concluded that summer afternoon precipitation frequencies increased by 30 percent to 60 percent in the Southwest and decreased by 15 percent to 30 percent in the Southeast. In winter, diurnal amplitude in precipitation frequencies increased in the Southwest by 30 percent to 47 percent, while in the Northwest, it decreased a similar amount.

In an apparent contrast with the Karl and Knight (1998) findings, Kunkel et al. (1999) examined short duration events (weekly or less) for the United States and Canada since 1931. They note that extended periods of below-average events occurred in the 1930s and 1950s with above average periods during the 1940s, the 1980s, and the 1990s. Since 1931, the linear trend has been positive and significant for the southwestern United States, the central Great Plains, and across the middle Mississippi River and southern Great Lakes basins. In the midwestern United States, for example, a 101-year record from 1896 to 1996 showed that heavy event frequencies from 1896 to 1906 were higher for all decadal periods except 1986 to 1996. They argued that "interpretation of the recent upward trends must account for the possibility of significant natural forcing of variability on century timescales" (Kunkel et al. 1999, 2515). Moreover, they concluded, "attribution of the cause of the recent upward trend should consider potential natural forcing factors that may make a significant contribution. There is no implication in these results that the upward trends will necessarily continue" (Kunkel et al. 1999, 2526).

More recently, however, Kunkel et al. (2003) revisited the analysis of Kunkel et al. (1999) with a new and extended dataset that provided extensive quality control and statistical tests to ensure confidence in the early, data sparse portions of the record. They found that the frequency of heavy precipitation was high during the late nineteenth and early twentieth centuries, reached a minimum in the 1920s and 1930s, and gradually increased to the late twentieth century. They concluded that "the frequencies at the beginning of the twentieth century were nearly as high as during the late twentieth century . . . suggesting that natural variability cannot be discounted as an important contributor to the recent high values" (1900).

The apparent discrepancy between the results of Karl and Knight (1998) and those of Kunkel et al. (1999, 2003) can be easily explained. Based on the results of Kunkel et al. (2003), an analysis that extends from 1910 to 1996 (i.e., that of Karl and Knight) misses the high precipitation frequencies that occurred during the late nineteenth and early twentieth centuries and, instead, begins during the low frequency period during the 1920s and 1930s. Both studies agree that precipitation frequencies since about 1920 have been steadily increasing. However, the extended temporal analysis of Kunkel et al. (2003) argues strongly that this upward trend may not be of anthropogenic origin but simply a result of a long-term natural cycle.

As previously demonstrated, the inability of climate models to simulate the observed precipitation seriously undermines the results from climate model prognostications. A further flaw in climate model simulations of precipitation lies in their inability to replicate the observed spatial and interannual variability. Soden (2000) demonstrated, for example, that while different models vary considerably with respect to the simulation of tropical precipitation, the interannual variability in model-simulated precipitation is nearly an order of magnitude less than the observed variability (figure 6.4). With regard to figure 6.4, Soden comments:

> The chief feature [in the figure] . . . is the substantial difference between the observed and GCM-simulated variation in tropical-mean precipitation. . . . Thus, not only do the GCMs differ with respect to the observations, but the models also lack coherence among themselves. It is noted, however, [t]hat even the extreme models exhibit markedly less precipitation variability than observed. . . . Moreover, the diversity of models considered here requires that this error not be sensitive to the differing physical parameterizations between existing GCMs. Rather, if the GCMs are in error, this deficiency would presumably reflect a more fundamental flaw common to all models. (541–42)

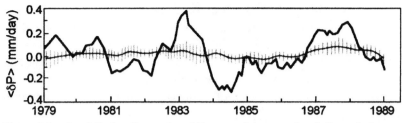

Figure 6.4. Precipitation time series of the mean interannual variations of observations (thick solid line) and multimodel ensemble mean GCM simulations (thin solid line) for the tropics (adapted from Soden 2000). The range bounded by one intermodel standard deviation is represented by the vertical lines centered on the GCM ensemble mean.

It is difficult to argue for an enhanced hydrologic cycle when climate models are unable to simulate correctly the observed interannual variability in precipitation. But more important, Soden (2000) recognizes that this difference in observations and model simulations occurs despite varied physical model parameterizations. Thus, assessments of changes in the temporal distribution of precipitation as made by climate models are highly unreliable.

Changes in Flood and Drought Frequencies

Over the twentieth century (1900 to 1995), there were relatively small increases in global land areas experiencing severe drought or severe wetness. In many regions, these changes are dominated by inter-decadal and multi-decadal climate variability, such as the shift in ENSO towards more warm events. In some regions, such as parts of Asia and Africa, the frequency and intensity of droughts have been observed to increase in recent decades.

—IPCC 2001, 5

Recall that flood frequencies are affected not only by trends in rainfall, but also by changes in land use, urbanization, and channelization. Thus, it is difficult to attribute trends in stream flow to climatic changes. For example, two major studies that have examined stream flow trends in the United States have yielded divergent results. Lins and Slack (1999), using a non-parametric statistical test, examined percentiles of stream flow for the twentieth century and found the most prevalent increases in the minimum and medium percentiles and least prevalent in the maximum percentiles. Decreases in stream flow occurred in parts of the Pacific Northwest and the southeastern United States. They concluded, "Two general patterns emerge; trends are most prevalent in the annual minimum to median flow categories and least prevalent in the annual maximum category. . . . Hydrologically, these results indicate that the conterminous US is getting wetter, but less extreme" (227).

By contrast, Groisman et al. (2001) examined stream flow in the United States and found significant increases, particularly for the highest flow events and months with the highest flows, which is consistent with their analysis of increases in extreme precipitation. Stream flow increases were most significant for the eastern half of the United States, although the western half has shown no increases in peak stream flow due to decreases in snow cover. But these apparent discrepancies are relatively easy to explain. The two studies in fact answered different questions. Lins and Slack (1999) addressed the question "Are trends occurring in stream flow percentiles?" whereas Groisman et al. (2001) asked, "Of the total volume of water that changed, how much of that water came from a specific percentile?" (Lins 2003). Although the percentiles representing the largest stream flow had the smallest percentage increase (by far), that small percentage increase amounts to a considerable volume of water, since the annual maximum flow can be two or three orders of magnitude larger than the annual minimum flow. Thus, the conclusions of Groisman et al. (2001) appear to largely result from the extreme skew in the stream flow distribution.

In a more recent article, Labat et al. (2004) argue that, based on an analysis of 221 rivers from around the world for 1925 to 1994, an increase of 1°C results in an increase in the global runoff of 4 percent. Unfortunately, their analysis focused on large rivers, without addressing the impact of urbanization on their runoff. But more important, their analysis is affected by a single outlier (the 1925 datum); if that point is removed, the results are no longer statistical significant. Thus the conclusions of Labat et al. appear to be flawed (see also Legates et al. 2005).

With respect to drought and moisture surplus over global land areas, Dai et al. (1998) used the Palmer Drought Severity Index (PDSI) and found large multiyear and decadal variations in both drought and moisture surplus. Although they noted that the trends were small, increases in drought frequencies were observed over the Sahel, eastern Asia, Southern Africa, the United States and Europe, while increases in moisture surplus were observed in the United States, and Europe as well. But the use of the PDSI is not without problems. The PDSI is based on a relative index, using a simplistic water balance model with uniform soil and land use properties across large climatic regions. It is a relative measure, in that drought and moisture surplus frequencies are standardized for each location—PDSI values for the Texas Panhandle and eastern North Carolina can be identical even though moisture conditions are extremely different. Moreover, the PDSI is not well suited to assess longer-term hydrologic impacts, owing to its inherent annual timescale and lack of a snow budget (all precipitation is treated as rainfall). Runoff, and hence stream flow response, is underestimated as a result of a lack of a proper treatment of snowfall, frozen soil conditions, and the delay between precipitation and runoff. Guttman (1998) and Hayes et al. (1999) have documented the limitations in the PDSI and have suggested the use of the standardized precipitation index (SPI) instead. No global assessment of changes in the SPI is yet available, however.

In general, this discussion serves to underscore the difficulties in interpreting stream flow trends as changes in the climate. As previously stated, stream flow is affected by changes in land use, urbanization, and channelization, as well as precipitation. The large changes that have occurred in the United States resulting from urban growth and efforts of the U.S. Army Corps of Engineers substantially affect stream flow and undermine efforts to detect climate change signals.

Given the difficulty in simulating precipitation with a GCM, model projections of drought, floods, and changes in stream flow are tenuous at best. Nevertheless, the IPCC (2001) argues that the risk of drought has likely increased in a few areas and this risk will become greater in the future, particularly in the continental interiors (table 6.1). Given the contradictory

Table 6.1. Confidence estimates for changes in several hydrological variables/events (adapted from IPCC 2001, 15).

Changes in Climatic Event	Confidence in Observations (latter half of the 20th century)	Confidence in Projections (during the 21st century)
Precipitation Increases	Likely, over many Northern Hemisphere mid- to high-latitude land areas. Insufficient data or conflicting analyses exist for other areas.	Very likely, over many areas.
Increased summer continental drying and drought risk	Likely, in a few areas	Likely, over most mid-latitude continental interiors. (Lack of consistent model projections in other areas)
Increase in tropical cyclone mean and peak precipitation intensities	Insufficient data for assessment	Likely, over some areas
Changes in tropical cyclone location and frequency	Uncertain	Uncertain

assessments of current trends in floods, droughts, and stream flow and the lack of accuracy in model projections of present-day precipitation, let alone the questionable changes in precipitation, assertions of changes in the frequencies of these variables resulting from climate changes cannot be relied on as fact.

Changes in Storm Frequencies in the Tropics and Extra-tropics

> Changes globally in tropical and extra-tropical storm intensity and frequency are dominated by inter-decadal to multi-decadal variations, with no significant trends evident over the twentieth century. Conflicting analyses make it difficult to draw definitive conclusions about changes in storm activity, especially in the extra-tropics.
>
> —IPCC 2001, 13

This statement by the IPCC regarding tropical storm events is consistent with the results of Easterling et al. (2000) and, in particular, Goldenberg et al. (2001, 477), who concluded "there have been various studies investigating the potential effect of long-term global warming on the number and strength of Atlantic-basin hurricanes. The results are inconclusive." Henderson-Sellers et al. (1998, 19) earlier had argued that "in the two regions where reasonably reliable records exist (the North Atlantic and the western North Pacific), substantial multidecadal variability (particularly for intense Atlantic hurricanes) is found, but there is no clear evidence of long-term trends." Research by Krishnamurti et al. (1998), Walsh and Ryan (1999), and Meehl et al. (2000) have suggested that a warmer world would mean an increase in tropical storm activity, owing largely to the increased water vapor content of the atmosphere, while Bengtsson et al. (1996) and Yoshimura et al. (1999) argue for a decrease in tropical storm activity. Clearly, there is no consensus on the issue.

Henderson-Sellers et al. (1998, 19) further argued, "It is emphasized that the popular belief that the region of cyclogenesis will expand with the 26°C [sea surface temperature] isotherm is a fallacy. The very modest available evidence points to an expectation of little or no change in global frequency." Because tropical cyclone formation is often linked to sea surface temperatures exceeding 26°C (78.8°F), it had been speculated that an increase in the area of warmer waters would increase tropical cyclone frequencies and intensities (Ryan et al. 1992). Despite these caveats, the *United States National Assessment,* a report that relied heavily on the TAR, nevertheless erroneously claims, "While it is not yet clear how the numbers and tracks of hurricanes will change, projections are that peak wind speed and rainfall intensity are likely to rise significantly" (USNA 2001a, 16) and "increases in hurricane wind strength could result from future elevated sea surface temperatures over the next 50 to 100 years" (USNA 2001a, 468). Citing results from Knutson et al. (1998) and Kerr (1999), the *National Assessment* suggests a 25 percent increase in the destructive power of a hurricane. However, this is directly at odds with chapter 1 of the same assessment, where the studies of Bengtsson et al. (1996) and Yoshimura et al. (1999) indicate a decreased tropical storm frequency (59).

With the suggestion by the IPCC that there will be a shift in ENSO to more warm (El Niño) events, we might expect a decrease in landfalling hurricanes in the United States with an increase in Western Pacific hurricanes. Bove et al. (1998), for example, have suggested that the probability of a landfalling hurricane decreases during warm (El Niño) episodes while cold (La Niña) events enhance the probability of a hurricane strike. Knutson et al. (1998) and Knutson and Tuleya (1999) argue for increased hurricane activity in the western Pacific during warm (El Niño) events. Meehl et al. (2000), however, conclude that making future projections about ENSO events and its impact on tropical storm frequencies is complicated.

With respect to the extra-tropics, Hayden (1999, 1396) wrote, "There has been no trend in North America-wide storminess or in storm frequency variability found in the record of storm tracks for the period 1885–1996. . . . It is not possible, at this time, to attribute the large regional changes in storm climate to elevated atmospheric carbon dioxide." Hayden goes on to say, "[Model] projections of North American storminess shows no sensitivity to elevated carbon dioxide. It would appear that statements about storminess based on [model] output statistics are unwarranted at this time" and that "It should also be clear that little can or should be said about change in variability of storminess in future, carbon dioxide enriched years." However, Sinclair and Watterson (1999) argued:

> Doubled CO_2 leads to a marked decrease in the occurrence of intense storms [in the extratropics] . . . one exception is over the South Pacific where there is a suggestion of an increased incidence of cyclones at the intense end of the spectrum. Reductions in average cyclone central pressure that have been used in other studies to promote the possibility of enhanced storminess under greenhouse warming, are more likely the result of global-scale sea level pressure falls rather than any real increase in cyclone circulation strength.

Much of the problem regarding storminess, as predicted by model-based analyses, centers on the model representation of topography. In spectral climate models, such as the Canadian Climate Centre, GFDL, and Hadley GCMs, topography is represented as a series of wave functions (figure 6.5). Lindberg and Broccoli (1996) clearly demonstrated that such representation results in extreme mountains and valleys being produced over the relatively flat oceans. Fyfe and Flato (1999) also demonstrated the limitations of model representations of the Rocky Mountains. More recently and using a higher resolution model, Biasutti et al. (2003) has shown that along the coastlines, extreme topographical variations are produced by the Gibbs phenomenon—wildly varying undulations produced in areas where extreme relief transitions into flat areas, such as the oceans (figure 6.6). This provides excessive topographic relief that adversely affects storminess assessments as well as precipitation simulations.

With respect to global-scale drops in sea-level pressure, their comment refers to the failure of some models to conserve mass, thereby producing global reductions in sea-level pressure. Sinclair and Watterson (1999) had noted that some studies had focused on decreases in sea-level pressure over time as an indicator of increased storminess, although they had failed to ac-

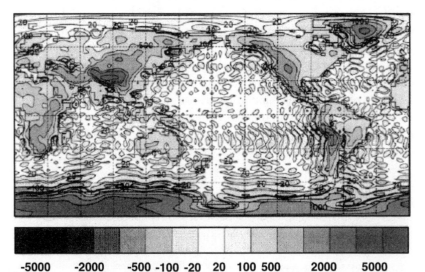

-5000   -2000   -500 -100 -20   20   100 500   2000   5000

Figure 6.5. Surface elevation as represented in a spectrally based general circulation model (GCM) using a standard R30 truncation (Lindberg and Broccoli 1996). Note the extensive topographical forcing over the oceans and the simplistic representation of the Rocky Mountains (see also Fyfe and Flato 1999).

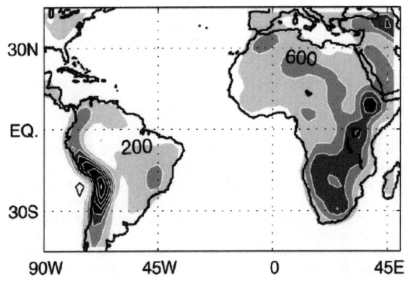

Figure 6.6. Surface elevation over Africa, South America, and the central Atlantic Ocean as represented in a spectrally based general circulation model (GCM) using a standard T42 truncation scheme (Biasutti et al. 2003). The contour interval is 400m. Negative altitudes of -200m occur west of the Andes while the highest peak in the Andes is only 3,000m high.

count for the fact that the climate model did not respect the Law of Conservation of Mass.

Overall, the *Third Assessment Report* presents a picture of increased storminess that is less certain than that presented in the *Second Assessment Report* (IPCC 1996). Although the *Second Assessment Report* indicated that evidence for changes in extra-tropical storminess is inconclusive, it did contend that there was "some evidence of change" in selected areas (IPCC 1996, 171). With respect to tropical cyclones, the *Second Assessment Report* expressed doubts about the quality and consistency of the data used to assess changes in tropical cyclone frequencies and intensities. Despite the more detailed analyses conducted since 1996, arguments about changes in tropical storms remain inconclusive. That also holds for changes in thunderstorm frequencies and hail occurrences (Changnon and Changnon 2000; Dai 2001a, b) as well as tornado frequencies, including the occurrence of "significant" tornadoes (F3 and higher).

Surprisingly and in contradiction to the IPCC Scientific Assessment (2001), the United States National Assessment (2001a, 16) asserts, "It is likely that the observed trends toward an intensification of precipitation events will continue. Thunderstorm and other intensive rain events are likely to produce larger rainfall totals. Projections [for hurricanes] are that peak wind speed and rainfall intensity are likely to rise significantly."

Yet clearly the science does not back up that statement.

CONCLUSION

The IPCC's *Second Assessment Report* was among the first to claim that an enhanced hydrologic cycle—more extremes in floods, droughts, heavy rainfall, and storminess—would result from rising global air temperatures. The *Third Assessment Report's* Summary for Policymakers goes further in its statements, although that summary—which is the only part many journalists and politicians actually read—often is in direct contrast with the scientific report that accompanies it. Among the main problems in assessing such claims is that biases in precipitation measurement, owing to both the spatial limitations in precipitation gauge networks and deficiencies in can-type gauges, make trend assessments difficult. Moreover, anthropogenic, but nonclimatic, effects (such as urbanization, deforestation, and channelization) and increasing demands for water from a growing population prejudice our interpretation and perception of floods and droughts.

It would seem, therefore, that General Circulation Models (GCMs) would provide the evidence necessary to answer such questions, since they are not prone to the gauge measurement bias concerns and land surface changes can be held constant. Alas, these models do not adequately represent the hydrologic cycle to a degree sufficient to allow an assessment of possible changes in the precipitation climate. GCMs are limited by (1) our incomplete understanding of the climate system and its interactions, (2) coarse spatial and temporal resolutions, (3) the inability of the model to reproduce many phenomena that are vital to assessing variability in the hydrologic cycle (e.g., most forms of severe weather), and (4) the interconnected nature of the climate system, which leads to inaccuracies in virtually all aspects of climate adversely affecting the simulation of the hydrologic cycle, and vice versa. Even the use of statistical downscaling techniques and nested grid models does not alleviate the problems, since it is the GCM that drives the large-scale forcing on which such techniques are based.

With respect to changes in precipitation totals and amounts, slight increases have been observed over the past several decades. But those increases are far below the interannual variability and the biases and uncertainties associated with the measurement process. Changes in precipitation frequency and intensity appear to have increased slightly over the past several decades. That finding is consistent with the *Third Scientific Assessment Report* (IPCC 2001), although recent results by Kunkel et al. (2003) argue against an anthropogenic cause. Have flood and drought frequencies changed? The answer appears to be no, which is consistent with the *Third Scientific Assessment Report* (IPCC 2001). Finally, no significant changes in tropical and extratropical storm frequencies have occurred, which is again consistent with the *Third Scientific Assessment Report* (IPCC 2001). Interestingly, therefore, it is only the IPCC Summary for Policymakers (IPCC 2001), and not the IPCC Scientific Assessment itself, that argues for an enhanced hydrologic cycle. "Global warming," it summarizes, "is likely to lead to greater extremes of drying and heavy rainfall and increase the risk of droughts and floods that occur . . . in many different regions."

Obviously, however, that claim directly contradicts the scientific evidence—including evidence that the IPCC itself presents.

NOTE

1. In SI units, ρ_w equals 1025 kg m^{-3}, L equals 2.5×10^6 J kg^{-1}, g equals 9.8 m s^{-2}, C_p equals 1004.67 J kg^{-1} K^{-1}, and p equals 0.8×101325 N m^{-2}.

REFERENCES

Akinremi, O. O., S. M. McGinn, and H. W. Cutforth. 1999. Precipitation trends on the Canadian Prairies. *J Climate* 12:2996–3003.

Bengtsson, L., M. Botzet, and M. Esch. 1996. Will greenhouse gas-induced warming over the next 50 years lead to higher frequency and great intensity hurricanes? *Tellus* 48A:57–73.

Biasutti, M., D. S. Battisti, and E. S. Sarachik. 2003. The annual cycle over the tropical Atlantic, South America, and Africa. *J Climate* 16(15):2491–2508.

Bove, M. C., J. B. Elsner, C. W. Landsea, X. F. Niu, and J. J. O'Brien. 1998. Effect of El Niño on US landfalling hurricanes, revisited. *Bull Amer Met Soc* 79(11): 2477–82.

Changnon, S. A., and D. Changnon. 2000. Long-term fluctuations in hail incidences in the United States. *J Climate* 13(4): 658–64.

Dai, A. 1999. Recent changes in the diurnal cycle of precipitation over the United States. *Geophys Res Lett* 26(3):341–44.

Dai, A. 2001a. Global precipitation and thunderstorm frequencies. Part I: Seasonal and interannual variations. *J Climate* 14(6):1092–1111.

Dai, A. 2001b. Global precipitation and thunderstorm frequencies. Part II: Diurnal variations. *J Climate* 14(6):1112–28.

Dai, A., K. E. Trenberth, and T. R. Karl. 1998. Global variations in droughts and wet spells: 1900–1995. *Geophys Res Lett* 25(17):3367–70.

Doherty, R., and L. O. Mearns. 2000. A comparison of simulations of current climate from two coupled atmosphere—ocean global climate models against observations and evaluation of their future climates: Report in support of the National Assessment. National Center for Atmospheric Research, Boulder, Colorado.

Easterling, D. R., J. L. Evans, P. Ya. Groisman, T. R. Karl, K. E. Kunkel, and P. Ambenj. 2000. Observed variability and trends in extreme climate events: A brief review. *Bull Amer Met Soc* 81(3):417–25.

Eishied, J. K., H. F. Diaz, R. S. Bradley, and P. D. Jones. 1991. A comprehensive precipitation data set for global land areas. U.S. Department of Energy Monograph TR051, DOE/ER-69017T-H1, Washington, D.C.

Felzer, B., and P. Heard. 1999. Precipitation differences amongst GCMs used for the U.S. National Assessment. *J Amer Water Res Assoc* 35(6):1327–40.

Fyfe, J. C., and G. M. Flato. 1999. Enhanced climate change and its detection over the Rocky Mountains. *J Climate* 12(1):230–43.

Goldenberg, S. B., C. W. Landsea, A. M. Mestas-Nuñez, and W. M. Gray. 2001. The recent increase in Atlantic hurricane activity: Causes and implications. *Science* 293:474–79.

Groisman, P. Ya., and D. R. Legates. 1994. Accuracy of historical United States precipitation data. *Bull Amer Met Soc* 75(2):215–27.

Groisman, P. Ya., and D. R. Legates. 1995. Documenting and detecting long-term precipitation trends: Where we are and what should be done. *Clim Change* 31: 601–22.

Groisman, P. Ya., R. W. Knight, and T. R. Karl. 2001. Heavy precipitation and high stream flow in the contiguous United States: Trends in the twentieth century. *Bull Amer Met Soc* 82(2):219–46.

Groisman, P. Ya., T. R. Karl, D. R. Easterling, R. W. Knight, P. F. Jamason, K. J. Hennessy, R. Suppiah, et al. 1999. Changes in the probability of heavy precipitation: Important indicators of climatic change. *Climatic Change* 42:243–83.

Guttman, N. B. 1998. Comparing the Palmer Drought Index and the standardized precipitation index. *J Amer Water Res Assoc* 34(1):113–21.

Hayden, B. P. 1999. Climate change and extratropical storminess in the United States: An assessment. *J Amer Water Res Assoc* 35(6):1387–98.

Hayes, M. J., M. D. Svoboda, D. A. Wilhite, and O. V. Vanyarkho. 1999. Monitoring the 1996 drought using the standardized precipitation index. *Bull Amer Met Soc* 80(3):429–38.

Henderson-Sellers, A., H. Zhang, G. Berz, K. Emanuel, W. Gray, C. Landsea, G. Holland, et al. 1998. Tropical cyclones and global climate change: A post-IPCC assessment. *Bull Amer Met Soc* 79(1):19–38.

IPCC. *Climate Change 1995: The science of climate change.* Contribution of Working Group I to the Second Assessment Report of the Intergovernmental Panel on Climate Change. Cambridge: Cambridge University Press, 1996.

IPCC. *Climate change 2001: The scientific basis.* Contribution of Working Group I to the Third Assessment Report of the Intergovernmental Panel on Climate Change. Cambridge: Cambridge University Press, 2001.

Karl, T. R., and R. W. Knight. 1998. Secular trends of precipitation amount, frequency, and intensity in the United States. *Bull Amer Met Soc* 79(2):231–41.

Karl, T. R., R. W. Knight, D. R. Easterling, and R. G. Quayle. 1996. Indices of climate change for the United States. *Bull Amer Met Soc* 77(3):279–92.

Kerr, E. A. 1999. Thermodynamic control of hurricane intensity. *Nature* 401:665–69.

Knutson, T. R., and R. E. Tuleya. 1999. Increased hurricane intensities with CO_2-induced warming as simulated using the GFDL hurricane prediction system. *Clim Dynamics* 15:503–19.

Knutson, T. R., R. E. Tuleya, and Y. Kurihara. 1998. Simulated increase of hurricane intensities in a CO_2 warmed climate. *Science* 279:1018–20.

Krishnamurti, T. N., R. Correa-Torres, M. Latif, and G. Daughenbaugh. 1998. The impact of current and possibly future sea surface temperature anomalies on the frequency of Atlantic hurricanes. *Tellus* 50A:186–210.

Kunkel, K. E., K. Andsager, and D. R. Easterling. 1999. Long-term trends in extreme precipitation events over the conterminous United States and Canada. *J Climate* 12(8):2515–27.

Kunkel, K. E., D. R. Easterling, K. Redmond, and K. Hubbard. 2003. Temporal variations of extreme precipitation events in the United States: 1895–2000. *Geophys Res Lett* 30(17):1900–1903.

Labat, D., Y. Godderis, J. L. Probst, and J. L. Guyot. 2004. Evidence for global runoff increase related to climate warming. *Adv Water Res* 27:631–42.

Legates, D. R. 1987. A climatology of global precipitation. *Publ in Climatol* 40(1).

Legates, D. R. 1995. Precipitation measurement biases and climate change detection. Presented to the Sixth Symposium on Global Change Studies. American Meteorological Society, Dallas.

Legates, D. R., and T. L. DeLiberty. 1993. Precipitation measurement biases in the United States. *Water Res Bull* 29:855–61.

Legates, D. R., and C. J. Willmott. 1990. Mean seasonal and spatial variability in gauge-corrected, global precipitation. *Int J Clim* 10(2):111–27.

Legates, D. R., H. F. Lins, and G. J. McCabe. 2005. Comments on "Evidence for global runoff increase related to climate warming," by Labat et al. *Adv Water Res,* in press.

Lindberg, C., and A. J. Broccoli. 1996. Representation of topography in spectral climate models and its effect on simulated precipitation. *J Climate* 9(11):264–59.

Lins, H. F. 2003. Personal communication. United States Geological Survey, Reston, Va.

Lins, H. F., and J. R. Slack. 1999. Stream flow trends in the United States. *Geophys Res Lett* 26(2):227–30.

Meehl, G. A., F. Zwiers, J. Evans, T. Knutson, L. Mearns, and P. Whetton. 2000. Trends in extreme weather and climate events: Issues related to modeling extremes in projections of future climate change. *Bull Amer Met Soc* 81:427–36.

Mekis, E., and W. D. Hogg. 1999. Rehabilitation and analysis of Canadian daily precipitation time series. *Atmos-Ocean* 37(1):53–85.

New, M., M. Hulme, and P. D. Jones. 2000. Representing twentieth-century space-time climate variability. II: Development of 1901–1996 monthly grids of terrestrial surface climate. *J Climate* 13(13):2217–38.

New, M., M. Hulme, and P. D. Jones. 2001. Precipitation measurements and trends in the twentieth century. *Int J Clim* 21(15):1899–22.

Ryan, B. F., I. G. Watterson, and J. L. Evans. 1992. Tropical cyclone frequencies inferred from Gray's Yearly Genesis Parameter: Validation of GCM tropical climates. *Geophys Res Lett* 19:1831–34.

Sinclair, M. R., and I. G. Watterson. 1999. Objective assessment of extratropical weather systems in simulated climates. *J Climate* 12:3467–85.

Soden, B. J. 2000. The sensitivity of the tropical hydrological cycle to ENSO. *J Climate* 13(3):538–49.

Trenberth, K. E., A. G. Dai, R. M. Rasmussen, and D. B. Parsons. 2003. The changing character of precipitation. *Bull Amer Met Soc* 84(9):1205–17.

U.S. National Assessment. 2001a. Chapter 1, Scenarios for climate variability and change, 13–71.

U.S. National Assessment. 2001b. Chapter 16, Potential consequences of climate variability and change on coastal areas and marine resources, 461–87.

Walsh, K. J. E., and B. F. Ryan. 1999. Idealized vortex studies of the effect of climate change on tropical cyclone intensities. Presented to the Twenty-third Conference on Hurricanes and Tropical Meteorology, American Meteorological Society, Boston.

Yoshimura, J., M. Sugi, and A. Noda. 1999. Influence of greenhouse warming on tropical cyclone frequency simulated by a high-resolution AGCM. Presented to the Tenth Symposium on Global Change Studies, American Meteorological Society, Boston.

Zhang, X., L. A. Vincent, W. D. Hogg, and A. Niitsoo. 2000. Temperature and precipitation trends in Canada during the twentieth century. *Atmos-Ocean* 38:395–429.

7

PREDICTIVE SKILL OF THE EL NIÑO–SOUTHERN OSCILLATION AND RELATED ATMOSPHERIC TELECONNECTIONS

Oliver W. Frauenfeld

One of the most important sources of predictive skill for seasonal, interannual, and decadal climate variability and prediction is the El Niño–Southern Oscillation (ENSO). El Niño is an anomalous phase of the cyclical ocean–atmosphere system in the tropical Pacific. It has important consequences for weather and climate in not only the Pacific/North America sector but also the Northern Hemisphere, and perhaps the globe as a whole. The Southern Oscillation is a global-scale barometric oscillation with an average cycle of four years. The ENSO phenomenon certainly has a significant impact on climate. The global climate system is extremely complex, however, with numerous other important sources of climate variability, so evaluating the exact degree to which ENSO factors into the climate system remains a challenge despite decades of research.

Indeed, ENSO itself involves a complex series of interactions between tropical Pacific sea-surface temperatures (SSTs), changes in tropical convection due to ENSO-related SST anomalies, and the overlying tropical atmosphere. It is still unknown what triggers ENSO events, whether changes in the tropical atmosphere—the distribution of heat and energy—induces El Niño SST anomalies, or whether it is the El Niño SST anomalies that induce changes in the overlying atmosphere. Whatever the cause, once initiated, ENSO can then, via teleconnections, impact remote locations around the globe. Any potential externally induced or internally generated anomaly then feeds back either positively or negatively on the ocean and the atmosphere. Given the relatively short observational record of weather and climate variables, establishing cause and effect using empirical data sources alone is difficult. Because of their controllable environment, general circulation models (GCMs) are useful for a numerical simulation of climate by which to evaluate various sources of forced and natural climate variability, as well as cause and effect in our complex climate system. With these computer-based simulations, it seems possible to estimate, for example, the extratropical response of the midlatitude atmosphere to forcing such as the tropical Pacific Ocean, specifically ENSO. Even so, to predict the impact of ENSO, GCMs must first be able to model ENSO itself. Therein lies one of the biggest challenges still facing the modeling community today.

COMMON ASSUMPTIONS ABOUT ENSO

The *Third Assessment Report* (TAR) of the U.N. Intergovernmental Panel on Climate Change (IPCC 2001) examines the possibility of a connection between projected changes in ENSO with anthropogenic global warming, articulating its position in statements such as the following:

- "Analyses of several global climate models indicate that as temperatures increase due to increased greenhouse gases, the Pacific climate will tend to resemble a more El Niño–like state."
- "A different coupled model . . . shows a La Niña–like response and yet another shows an initial La Niña–like pattern which becomes an El Niño–like pattern."

When characterizing these results in summary, however, the IPCC seems to support the idea that El Niño conditions will predominate, as in, "A majority of models show a mean El Niño–like response in the tropical Pacific."

As to the strength of ENSO, the IPCC is equivocal, acknowledging varied model findings: "Attempts to address this question (of ENSO strength) using climate models have again shown conflicting results, varying from slight decreases in amplitude . . . to a small increase in amplitude."

With regard to the connections between ENSO and other weather phenomena, however, the IPCC is far less equivocal, citing modeling results that

> indicate that future seasonal precipitation extremes associated with a given ENSO event are likely to be more intense due to the warmer, more El Niño–like mean base state in a future climate . . . also in association

with changes in the extratropical base state in a future warmer climate, the teleconnections may shift somewhat with an associated shift of precipitation and drought conditions in future ENSO events.

Overall, the IPCC seems to come down on the side of increased El Niño frequency with global warming, and increased weather variability associated with El Niños, as mitigated by connections between El Niño and other circulation features. At the same time, it finds no strong evidence of either an increase or a decrease in the magnitude of individual El Niños.

In fact, however, much of what has been previously assumed about the magnitude of the teleconnections between ENSO and worldwide weather phenomena must be reconsidered. In this chapter, I delineate the limitations of GCM prognostications regarding ENSO and the analogous atmospheric circulation of the Northern Hemisphere. I begin with a history of ENSO and a description of the observed physical processes that govern the phenomenon. The mechanism for the ocean–atmosphere interactions related to ENSO is discussed next, followed by a description of the resulting modes of atmospheric circulation. Given that our climate is changing both due to natural variability (such as solar changes, as documented by many authors, e.g., Lean and Rind 1998) and anthropogenic sources (such as greenhouse changes, again documented by many authors, e.g., Michaels et al. 2000), changes in oceanic and atmospheric circulation should be evident. Those trends are outlined briefly, followed by a description of the uncertainty still surrounding our understanding of the complex observed ocean–atmosphere interactions involved in ENSO and its related teleconnections.

The complexity and uncertainty surrounding our understanding of the climate system leads naturally to a description of the models' limitations surrounding ENSO. First, the ability to model ENSO itself is assessed, followed by a description of GCMs' simulation of ENSO-related teleconnections and modes of atmospheric circulation. If this chapter seems overly critical of the models' ability to accurately simulate ENSO and analogous climate, bear in mind that recognizing their limitations is of critical importance, a position that is consistent with the IPCC's own findings. Even so, the IPCC was ultimately overly optimistic concerning the state of this science, as this paper's discussion of ENSO and related atmospheric teleconnections demonstrates. Despite a clear lack of certainty, policymakers continue to assert a connection between El Niño and global warming without a full appreciation for the seriousness of the GCM experiments' limitations.

Observations

EL NIÑO–SOUTHERN OSCILLATION

One of the most important contributors to interannual variability in hemispheric and global climate, the El Niño–Southern Oscillation has been recognized for more than a century as a dominant mode of Pacific Ocean and indeed global oceanic and atmospheric variability. The atmospheric component of ENSO, the Southern Oscillation, was noticed as early as the end of the nineteenth century as atmospheric pressure fluctuations between the western tropical Pacific and eastern Indian Ocean and the southeastern tropical Pacific (Hildebrandsson 1897) (figure 7.1). Lockyer and Lockyer (1902, 1904) confirmed this global-scale barometric oscillation and determined its period to be approximately four years.

Termed the Southern Oscillation by Sir Gilbert Walker, it is characterized by low atmospheric sea-level pressure (SLP) in regions dominated by tropical convection, ascending air, and rainfall—such as the west Pacific Australia-Indonesia region. Anomalously low pressure in these areas of convection corresponds to the concurrent occurrence of anomalously high pres-

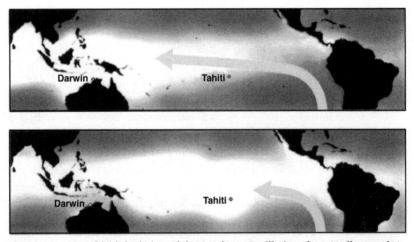

Figure 7.1.   Graphical depiction of the Southern Oscillation (from Wallace and Vogel 1994). The top panel depicts the positive phase (for November 1988) when pressure is higher in the vicinity of Tahiti than in the west Pacific, near Darwin, Australia. The east-west pressure difference along the equator causes enhanced northeast trade winds, as indicated by the long arrow. Conversely, during the negative phase (lower panel, November 1982) pressure rises in the west and falls in the east, which reduces or even reverses the pressure difference between Darwin and Tahiti, causing a weakening of the trade winds.

sure in the southeast Pacific—a region characterized by dry conditions and descending air. This phase of the Southern Oscillation sets up an East-West pressure gradient across the equatorial Pacific, which creates enhanced trade winds as well as enhanced mass exchange between the East and West Pacific (figure 7.1). The trade wind circulation constantly transfers air at low levels from subsidence regions—the subtropical deserts—to the intertropical convection regions, with the air being returned at tropospheric levels, thus completing a series of zonal circulation cells around the globe (Bjerknes 1969).

A number of indices have been devised to characterize the Southern Oscillation—the most common one being the Southern Oscillation Index (SOI), quantified by the standardized SLP difference between Darwin, Australia—representing SLP over the western Pacific—and Tahiti, French Polynesia—representing the eastern Pacific.

SST in the eastern equatorial Pacific is normally lower than its equatorial location would suggest, mainly due to the influence of a cold oceanic current flowing equatorward along the coast of Chile, and due to upwelling of cold deep water off the coast of Peru. Upwelling occurs also along the equatorial Pacific as the easterly trade winds generate an easterly equatorial current, which is deflected northward in the Northern Hemisphere and southward in the Southern Hemisphere, due to the Coriolis force and Ekman transport (figure 7.2). El Niño, the oceanic component of ENSO, is the occasional anomalous warming of the eastern equatorial Pacific, but is also commonly associated with a basin-scale warming extending from the West coast of South America to the International Dateline (Trenberth and Hoar 1996).

Like the Southern Oscillation, El Niño was known well before the turn of the twentieth century. Discovered by fishermen along the coasts of Peru and Ecuador, the phenomenon originally referred to the warm ocean current that typically appeared during the Christmas season and lasted for a few months (Wallace and Vogel 1994). In some years, however, people noticed that the oceanic surface waters in these regions were especially warm for an even longer period, thus interrupting the fishing season until late May or June. Over the years, these exceptionally long and intense events were termed El Niño, or "the baby boy," a reference to their Christmas birth. (El Niño has a "sister," known as La Niña, which occurs when the injection of cold water in the eastern equatorial Pacific becomes more intense than usual, resulting in anomalously cool surface waters in the eastern equatorial Pacific.) The suppression of this cold upwelling by an El Niño may in fact provoke the La Niña response because of the anomalous amount of un-upwelled cold water that is created by El Niño.

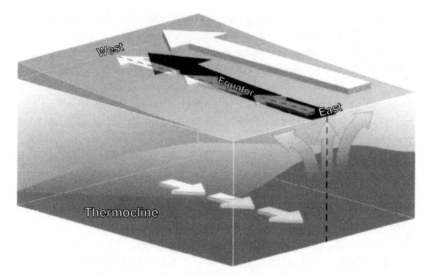

Figure 7.2.   In the equatorial Pacific, the easterly trade winds cause an easterly ocean current. Because of the Coriolis force, the current is deflected to the right in the Northern Hemisphere and to the left in the Southern Hemisphere, causing the surface waters to diverge, thus creating upwelling (upward arrows). The easterly trades also cause the surface waters to accumulate in the western Pacific. The thermocline, which marks the boundary between warm surface water and cold deep water, is tilted. It reaches almost up to the sea surface in the eastern equatorial Pacific (from Wallace and Vogel 1994).

During El Niño, the thermocline, a region of strong temperature gradient at the base of the well-mixed surface layer of the ocean, becomes depressed in the East Pacific, reducing equatorial upwelling and enhancing the eastward transport of warm surface waters from the western Pacific (figure 7.3). It is the location of the warm El Niño regions that determines the intensity and position of the convective regions, and thus the phase of the Southern Oscillation. Because El Niño often occur simultaneously with the low phase of the Southern Oscillation, the two phenomena are now collectively termed the El Niño—Southern Oscillation—a connection that was not made until the 1960s, although both the Southern Oscillation and El Niño were first noticed well over a hundred years ago (Bjerknes 1966, 1969).

OCEAN–ATMOSPHERE INTERACTIONS

Interactions between the ocean and the atmosphere are key to the simultaneous development of El Niño (SST) and Southern Oscillation (atmos-

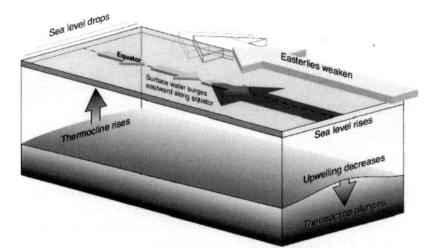

Figure 7.3.   During El Niño the easterly trades weaken, the thermocline "plunges" in the East Pacific, reducing equatorial upwelling and enhancing the eastward transport of warm surface waters from the western Pacific (from Wallace and Vogel 1994).

pheric/SLP) anomalies. The strong perturbation that ENSO provides to the tropical equatorial Pacific system has repercussions not only in the equatorial Pacific regions, but also in the extratropical Pacific, as well as in parts of the Northern Hemisphere and globe as a whole. It is these types of anomalies that the IPCC implies El Niño will change along with anthropogenic warming.

The areas of positive SST anomalies and corresponding convergence and convection not only alter east–west transport in the tropics, but also affect the poleward transport, and hence the exchange with extratropical regions. It was Bjerknes who first connected El Niño with the Southern Oscillation, and he who identified the teleconnections between ENSO and extratropical atmospheric anomalies during the winter season (Bjerknes 1969).

Bjerknes (1966, 1969) also provided a dynamic and thermodynamic mechanism for these teleconnections that today still serves as the basis for these interactions. During El Niño, or warm ENSO events, the east–west Walker circulation is weaker than normal while the north–south Hadley circulation is greatly enhanced. That intensified Hadley circulation transports increased momentum poleward to intensify midlatitude westerlies, thus impacting the Northern Hemisphere extratropics.

Bjerknes reasoned that as a warm ENSO episode develops, the West Pacific warm pool and associated heavy convection and precipitation extends eastward. The extensive air convergence into the sector must lead to upper tropospheric divergence not only by way of the zonal east–west Walker circulation, but also northward in the form of "Hadley outflow" (Rasmusson 1991). Temperate latitude variations in atmospheric quantities of the lower stratosphere, such as water vapor and total ozone, confirm the existence of enhanced Hadley circulation during ENSO (Angell 1981). The enhanced poleward transport leads in turn to a change in the North Pacific subtropical jet stream that can extend into the eastern Pacific or even the northern Gulf of Mexico. This feature is often associated with enhanced winter storminess in the Pacific Southwest and Gulf Coasts of the United States.

Similarly, the link between the tropics and the midlatitudes can also be found in large midlatitude circulations. During a warm ENSO event's mature phase, the major low latitude feature is an extensive pair of unusually strong high pressure cells—one to the north of the equator, characterized by clockwise flow, and another, counterclockwise circulation to the south. This in turn intensifies the difference in pressure between the poleward side of these high-pressure systems and the background environment, altering the normal midlatitude westerlies. This results in changes in the jet stream, especially an eastward extension of the East Asian subtropical jet (Bjerknes 1969). North and eastward of the jet stream extension is a teleconnection "triplet": a low-pressure circulation anomaly over the North Pacific, increased high pressure over western Canada, and another enhanced low-pressure anomaly over the eastern United States and the western Atlantic (Rasmusson 1991). Horel and Wallace (1981) point out that these three anomalies describe an apparent wave train pattern, which should force persistent weather anomalies normally associated with enhanced low and high pressure in the associated regions.

Theoretical results suggest that strong teleconnections to the midlatitudes are possible only when the westerlies extend from the middle latitudes into the equatorial troposphere over the vicinity of the heat source. For the Northern Hemisphere, this therefore applies mainly during the winter half of the year, when the westerlies and the jet stream reach their farthest southernmost displacement. As a result, the teleconnections to middle and high latitudes should occur primarily during the winter half of the year, as is observed with the aforementioned teleconnection "triplet" known as the Pacific/North America (PNA) teleconnection pattern (Horel and Wallace 1981; Wallace and Gutzler 1981) (figure 7.4).

Oceanic forcings have also been theorized that link the tropics and extratropics. In the aftermath of strong ENSO events, the tropical ocean can transmit responses to the North Pacific (Jacobs et al. 1994). When the easterly trade winds weaken during ENSO, an eastward-traveling equatorial

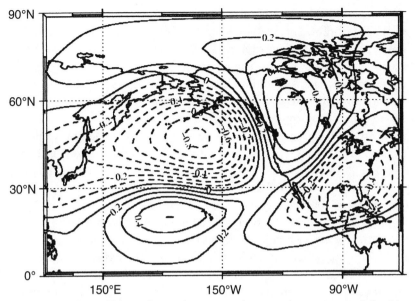

Figure 7.4. The Pacific/North America (PNA) teleconnection pattern as defined by Wallace and Gutzler (1981). The contours represent the correlations between their PNA index and the mid-tropospheric pressure field. Contour interval is 0.1; negative correlations are dashed, positive correlations solid.

"Kelvin" wave can be activated in the tropical ocean. When this wave impacts the eastern boundary of the Pacific, northward and southward traveling coastal Kelvin waves are generated, which can result in a periodic warming of the coastal waters in the extratropical Pacific (Emery and Hamilton 1985), including the North American coast.

Atmospheric Circulation Anomalies

PACIFIC SECTOR

The PNA pattern is one of the most prominent modes of year-to-year climate variability, especially during the Northern Hemisphere winter. Its positive phase is characterized by an anomalous southward extension of the westerlies in the vicinity of the Aleutian Low and the eastern United States and anomalous ridging (northward movement) of the westerlies over the Rockies (Yin 1994). This phase is therefore associated with a more north to south, or "meridional," upper-level atmospheric circulation, compared to the more normal "zonal," or west-to-east flow.

Conversely, the negative phase is characterized by zonal flow—a weakened Aleutian Low and less troughing over the southeastern United States, as well as a weakened ridge over the Rockies (Yin 1994). The positive phase associates with above-normal temperatures in the western United States, and drier than normal conditions in the Southeast (Yin 1994). Furthermore, the eastern–southeastern United States may experience intrusions of polar air and enhanced storminess. During the negative phase, the western United States can be characterized by relatively cold and wet conditions, while the eastern United States may experience dry and warm conditions (Yin 1994).

The PNA pattern has been shown to vary depending on the phase of ENSO (Arkin et al. 1980; Horel and Wallace 1981; Trenberth 1997; Frauenfeld and Davis 2000). Horel and Wallace (1981) correlated Northern Hemisphere winter circulation patterns at 700 mb (around 10,000 feet) with the Southern Oscillation Index and found that warm ENSO events are associated with (1) a substantial southward dip in these westerlies in a broad belt across the North Pacific, extending westward into Siberia, (2) a northward migration of them over western Canada, and (3) another southward movement over the southeastern United States. As is obvious, those anomalies strongly resemble the well-known PNA pattern, as well as another one known as the Western Pacific (WP) teleconnection pattern, which is defined as a primary mode of low-frequency variability over the North Pacific that, during winter and spring, consists of a north–south dipole, with one center over the Kamchatka Peninsula and another broad center of opposite sign over portions of southeastern Asia and the low latitudes of the extreme western North Pacific (in the summer and fall, a third center appears over Alaska and the Beaufort Sea, of opposite sign to the western North Pacific). Thus, a number of preferred atmospheric circulation modes exist that are potentially associated with ENSO's tropical forcing. Further, there are other patterns of Northern Hemisphere circulation, as identified by Wallace and Gutzler (1981), some of which are dominant in the Atlantic rather than the Pacific.

ATLANTIC SECTOR

In the Atlantic region, the major mode of atmospheric circulation variability is an oscillation between the Atlantic subtropical (Azores) high (whose westward extension is often called the "Bermuda high" in popular meteorology), and the subpolar (Icelandic) low-pressure system. This Atlantic north–south dipole combines parts of Wallace and Gutzler's East Atlantic and West Atlantic teleconnection patterns, and is called the

North Atlantic Oscillation (NAO). The opposing phases of the NAO manifest themselves in terms of temperature and precipitation anomalies from the eastern United States to Europe and the Middle East. The positive phase, when both the Icelandic Low and the Azores High are anomalously strong and hence the westerly flow across the Atlantic is enhanced, above-normal temperatures are observed in the eastern United States and across northern Europe, and below-normal temperatures in Greenland and across southern Europe and the Middle East (figure 7.5). The positive phase of the NAO also causes above-normal precipitation over northern Europe and Scandinavia, and below-normal precipitation over southern and central Europe (Lamb and Peppler 1991).

Recently it has been recognized that this seesaw between subpolar and subtropical sea level pressure is observed throughout the Northern Hemisphere, and is perhaps not limited solely to the Atlantic region. The NAO therefore may be a regional manifestation of a hemispheric-scale atmospheric circulation pattern, which has been termed the Arctic Oscillation (AO) (e.g., Thompson and Wallace 1998).

Similar to the NAO, the AO's positive phase is characterized by decreased surface pressure over the Arctic and increased surface pressure over the Northern Hemisphere subtropical oceans (figure 7.6). This positive phase of the AO tends to be associated with a stronger jet stream and strengthened northerly winds, which is linked with increased precipitation over Northern Europe and Alaska and decreased precipitation in Spain and California. Eurasia tends to be warmer than normal due to the strong vortex and resulting strong winds transferring maritime air masses onto the continents. In North America, conditions tend to be colder than normal in eastern Canada, and warmer than normal in central Canada. Furthermore, storm systems that develop over the oceans take a more northerly track than normal.

The negative phase of the AO occurs when Arctic pressure is higher than normal and pressure in the midlatitudes is lower than normal, resulting in a weakened north–south pressure gradient and weaker atmospheric flow, as velocity is proportional to the difference in pressure. The opposite weather conditions to those associated with the positive phase characterize the negative phase of the AO.

LONG-TERM CLIMATE TRENDS

ENSO, driven by coupled processes, is one of the most important sources of interannual variability in the natural climate system (Philander 1990).

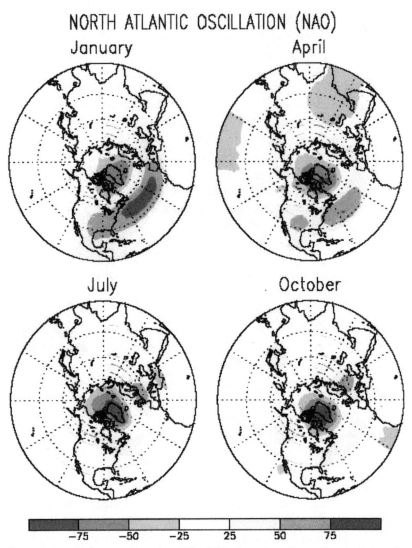

NORTH ATLANTIC OSCILLATION (NAO)
January   April

July   October

−75   −50   −25   25   50   75

Figure 7.5. Surface pressure anomaly fields for selected months, depicting the North Atlantic Oscillation's dipole between the Icelandic Low and the Azores High. Though the NAO is most prominent during winter (top left), it is in fact evident throughout the year (Source: NOAA Climate Prediction Center).

Although these events have an average frequency of two to seven years, recent decades have seen an apparent increase in the periodicity, and perhaps the intensity, of ENSO. Suddenly, in the winter of 1976–77, the tropical Pacific Ocean shifted toward a warm mode, which coincided with a global shift toward a warmer regime (Zhang et al. 1997). First

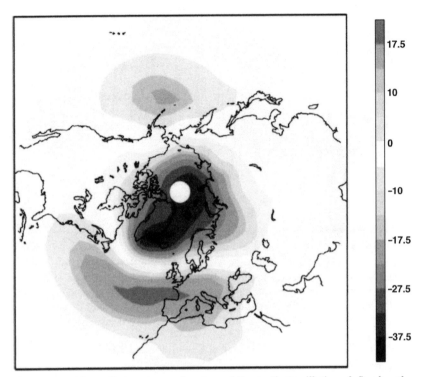

Figure 7.6. The surface pressure signature of the Arctic Oscillation, defined as the leading mode of the monthly Northern Hemisphere pressure anomalies (from Thompson and Wallace 2001).

noted by Quinn and Neal (1984, 1985) in the time series of ENSO indices and later in a more global context by Nitta and Yamada (1989), the "Pacific Climate Shift," as it is called (e.g., Miller et al. 1994), coincided with shifts in many climate records, such as North Pacific SSTs (Miller et al. 1994), oceanic heat content (Levitus et al. 2000), weather-balloon measured temperatures (Angell 1999), atmospheric sea level pressure and surface air temperature (Minobe 1997, 1999), the leading pattern of Pacific Ocean–Northern Hemisphere atmosphere interaction (Frauenfeld and Davis 2002; Frauenfeld 2003), and so forth (figure 7.7).

Since the great "Pacific Climate Shift" of 1977 there has been an apparent increase in the frequency of ENSO events, which has led to speculation that recent ENSO variability is unusual compared to past ENSO fluctuations (e.g., McPhaden 1999). Two major ENSO events, 1982–1983 and 1997–1998, occurred within fifteen years of each other, whereas usually such anomalous events occur only every thirty to forty years (Caron and

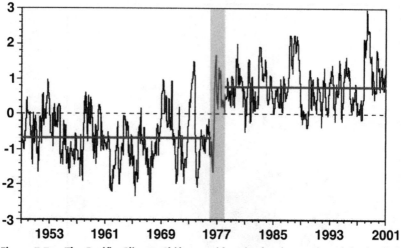

Figure 7.7. The Pacific Climate Shift, as evident in the time series of the leading pattern of Pacific Ocean–Northern Hemisphere atmosphere interaction of Frauenfeld and Davis (2002). The shaded box indicates the approximate timing of the shift, and the thick horizontal lines indicate the pre- and post-shift means.

O'Brien 1998). Furthermore, there was a prolonged ENSO event during the early 1990s.

Are such decadal changes in the frequency and intensity of ENSO linked with global warming, as some have suggested (IPCC, 2001)? Quantitative evidence, as provided here, demonstrates otherwise. Rather, such decadal fluctuations in both Pacific SSTs and atmospheric circulation are part of a multidecadal oscillation that involves ocean–atmosphere interactions containing no long-term signal that can be associated with greenhouse gas increases (Frauenfeld 2003; Frauenfeld and Davis 2002).

Like ENSO, these major modes of atmospheric circulation (PNA, NAO, AO) have exhibited long-term trends. Concurrent with the Pacific Climate Shift, changes in the PNA pattern occurred in the mid-1970s such that the Aleutian Low deepened and shifted eastward, resulting in increasing flow of warm and moist air over western North America and cooler and drier conditions over the central North Pacific (Hurrell 1996). During January 1977, dramatic stratospheric warming was observed, followed by the formation of a strong surface high-pressure system over the Arctic (Quiroz 1977). Subsequently, the NAO shifted into its positive phase around 1980, indicating anomalous low pressure near the Icelandic Low and high-pressure anomalies in the subtropical Atlantic in conjunction with stronger than normal westerlies over the midlatitude Atlantic Ocean.

Here is where the global warming question arises. That shift toward the NAO's positive phase has been correlated with the wintertime increases of surface air temperatures north of 20°N (Hurrell 1996). The shift has also been characterized as "unusual" anomalous atmospheric behavior, suggesting ties to ozone depletion and increases in greenhouse gases (IPCC 2001).

Yet that argument can only go so far: to the late 1990s, in fact. The NAO's time series was indeed increasing strongly during the mid-1970s through early 1990s, prompting researchers to proclaim a positive trend that seemed to them to coincide with rising greenhouse gas levels (e.g., Hurrell 1996). But then the NAO shifted, remaining negative throughout the late 1990s, and putting an end to this long-term trend, despite still increasing greenhouse gas levels (figure 7.8).

The overall linear trend may still be positive and the NAO is still being reported to exhibit "a strong positive trend" (Hoerling et al. 2001). But the time series itself is nonlinear and, especially in light of the NAO's negative departures during the late 1990s, such linear trend descriptions are as meaningless as the global warming implications they are purported to support.

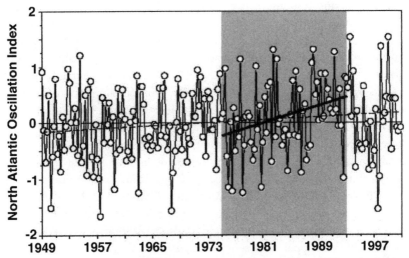

Figure 7.8.   Time series of the NOAA Climate Prediction Center's seasonal NAO index. During the 1970s to the early 1990s (shaded box), there was a strong positive trend, as indicated by the thick trend line. But the mid- to late 1990s were characterized by strong negative departures, resulting in an overall trend that is only slightly positive, driven primarily by the negative values throughout the 1950s and 1960s (dashed trend line).

The hemispheric manifestation of the NAO, the AO, has been significantly trending toward its positive phase, or a decrease in pressure over the polar region. That change in the AO is also characterized by a strengthening in the subpolar westerlies and a weakening of the jet stream in lower latitudes. As the pressure difference between the polar regions and the subtropics is closely linked with the Northern Hemisphere circumpolar vortex, these changes in the AO are also strongly evident in the vortex (Frauenfeld and Davis 2003). The biggest changes in the vortex are observed over Eurasia via a strengthened westerly flow and significant vortex contraction since approximately 1970. More than 50 percent of the observed warming trend over Eurasia, and 30 percent of the total extratropical wintertime Northern Hemisphere warming, has been attributed to these circulation changes (Thompson and Wallace 2001).

UNCERTAINTY ABOUT ENSO AND THE OTHER CIRCULATIONS

There is still substantial uncertainty as to the degree to which ENSO indeed induces these atmospheric circulation changes in the extratropics, whether these "responses" are actually different from noise, and whether other recurring response patterns to tropical forcing exist (Hoerling and Kumar 2002). Some observational evidence suggests that inter-ENSO variability is consistent with noise (Madden 1976), a notion that is supported by GCMs that find only weak extratropical responses to tropical Pacific SST forcing (Geisler et al. 1985; Kumar and Hoerling 1995). Yet other GCM results indicate that there is a strong extratropical signal in response to ENSO (Palmer and Owen 1986; Trenberth 1993). Even more uncertainty surrounds the impacts on extratropical climate from tropical non-ENSO forcing, as well as tropical forcing outside of the equatorial Pacific region (Palmer and Owen 1986; Hoskins and Sardeshmukh 1987; Kumar et al. 2001). Further GCM simulations show a significant extratropical signal in the PNA region during non-ENSO winters (Shukla et al. 2000) and a circulation response to ENSO forcing different from the traditional PNA teleconnection pattern in the midlatitude atmosphere (Straus and Shukla 2002). These different upper tropospheric height anomalies are evident in both observations (Mo and Livezey 1986; Trenberth et al. 1998) and model simulations (Lau 1985; Shukla and Wallace 1983; Peng et al. 2000).

Simple dynamic models predict a geographically fixed extratropical response, the "typical" PNA pattern, when driven by tropical heating (Webster 1981; Hoskins and Karoly 1981; Simmons 1982; Simmons et al. 1983;

Branstator 1990). There is thus significant uncertainty and debate about whether the dominant patterns of midlatitude atmospheric circulation in response to tropical forcing are related to preferred internal modes of variability; that is, whether ENSO merely shifts or amplifies preferred atmospheric circulation modes such as the PNA pattern, or whether tropical SST anomalies force distinct responses and are capable of generating new structures in the midlatitude atmosphere (Straus and Shukla 2002).

The nature of ENSO forcing and ocean–atmosphere interaction is complicated even further as any atmospheric response to tropical forcing feeds back on the ocean, and the ocean then in turn feeds back to the atmosphere. The atmosphere can therefore be argued to act as a bridge, spanning from the equatorial Pacific to the North Pacific, as well as to various other parts of the globe (Alexander et al. 2002). It is that "atmospheric bridge," a term first coined by Lau and Nath (1996), that contributes to the translation of tropical SST variability such as ENSO to the leading mode of SST variability on interannual to longer timescales in the North Pacific as well as other regions of the globe (Graham et al. 1994; Zhang et al. 1997; Gu and Philander 1997).

Weare et al. (1976) first noted the relationship between SST anomalies in the tropical Pacific and those of the North Pacific. Since then a number of studies have confirmed a dominant SST pattern in which the equatorial Pacific and the West coast of North America are characterized by anomalies of one sign, and the North Pacific is characterized by anomalies of the opposite sign (e.g., Deser and Blackmon 1995; Zhang et al. 1996). SST fluctuations related to ENSO are also evident in the Bering Sea (Niebauer 1984, 1988) and the South China Sea (Hanawa et al. 1989) during winter, and in the North Pacific during summer and fall (Reynolds and Rasmusson 1983; Wallace and Jiang 1987). These SST anomalies in regions distant from the tropics, but linked to the tropics, feed back to the overlying extratropical atmosphere.

Namias (1976) provides the first observational evidence and notes that in regions of negative SST anomalies, positive geopotential height anomalies, or ridges, form. These ridges in turn help generate a pool of warm water, which can then again feed back to the atmosphere and cause negative height anomalies, or troughs. Those feedbacks are especially evident in the extratropics, where the atmosphere has been found to induce changes in the ocean (Alexander 1992). Atmospheric forcing via surface energy fluxes, vertical turbulent mixing, and wind-driven vertical and horizontal motions can induce SST anomalies. Meridional temperature advection by anomalous ocean currents then plays an important role in the development of SST anomalies. Observational evidence exists that changes in atmospheric circulation lead to the formation of such SST anomalies in the North Pacific (Emery and Hamilton 1985). Davis (1976) provides statistical evidence that the correlations between the atmosphere and the extratropical ocean are strongest when the atmosphere leads the ocean by several months. Observational and modeling results also suggest the possibility that atmospheric circulation anomalies upstream of the Pacific, over Eurasia, affect the downstream variability over the Pacific (e.g., Barnett et al. 1989; Clark and Serreze 2000; Cohen and Entekhabi 1999; Frauenfeld and Davis 2002).

Obviously, the ocean–atmosphere interactions involving ENSO are extremely complex, and the potential teleconnections associated with ENSO in the Northern Hemisphere are even more complicated.

GCM SIMULATIONS OF ENSO
COUPLING TO OTHER CIRCULATIONS

The degree to which ENSO affects the global climate system, despite decades of research, is therefore still relatively uncertain, and significant controversy exists about the exact nature of tropical forcing on remote locations of the globe, and hence the actual impacts of ENSO in the Northern Hemisphere. Changes in both oceanic conditions and atmospheric circulation have occurred in recent decades, and the degree to which these changes are related to global climate change and global warming further complicates these issues. Angell (2000) points out that because ENSO is brought about by ocean–atmosphere interactions, it cannot actually be considered an *external* forcing mechanism on the atmosphere, since ENSO itself is subject to, and likely affected by, climate change.

The controlled environment of GCMs offers one way to attempt to sort out these issues. But capturing the global teleconnections associated with ENSO—the core of this issue—remains one of the biggest challenges facing current state-of-the-art GCMs that couple the ocean and the atmosphere (coupled GCMs, or CGCMs). Observed climate anomalies associated with ENSO differ substantially from one event to another, therefore part of the problem is the poor predictability of ENSO impacts. Further, even with a preferred atmospheric response attributed to ENSO, very small shifts in stationary wave positions of the atmosphere can make the difference between, for example, a wet rainy season and a dry one (Hoerling and Kumar 1997).

The climate responses involve a complex chain of events including local responses of tropical convection caused by anomalous SST forcing, the sensitivity of the midlatitudes' flow to the precipitation anomalies, and the exact structure of the circulation through which poleward energy propagation from tropical heat sources occurs. Because of those complexities, atmospheric general circulation models (AGCMs) are inadequate in that they cannot distinguish between variations of atmospheric circulation due to differences in SST anomalies (a "signal"), and variations that occur independently of anomalous SSTs ("noise") (Kumar and Hoerling 1997). GCMs may be capable of predicting tropical responses to SST forcing, but they cannot account for extratropical teleconnections (Hoerling et al. 1997).

MODELING TROPICAL SST VARIABILITY

The IPCC (2001) suggests that coupled models can provide credible simulations of both the present annual mean climate and the climatological seasonal cycle over broad continental scales for most variables of interest for climate change. That is a dubious assertion. In their summary of the El Niño Simulation Intercomparison Project, Latif et al. (2001) illustrate the exact opposite. They point out that in modeling mean *annual* SST in the tropical Pacific, most models have a cold bias and are *at least* 1°C too cold in the central equatorial Pacific, while others are *at least* 1°C too warm. Overall, the range of mean annual SST values among a suite of twenty-four CGCMs in the central equatorial Pacific—the key region for characterizing ENSO variability—is approximately 6°C (figure 7.9). Furthermore, most CGCMs show large errors in modeling SST near the boundaries of the Pacific. SSTs predicted for the eastern boundary, the region off Peru that is key in the onset of ENSO events, are much too high. SST conditions at the western boundary, the Australia-Indonesia region, tend to be modeled better, although some models also exhibit large errors in that region. Overall, the mean annual warm pool (El Niño) and cold tongue (La Niña) pattern in the equatorial Pacific is still a challenge for CGCMs (Latif et al. 2001).

If CGCMs cannot simulate the mean annual SSTs in the tropical Pacific, it is not surprising that accurately modeling the seasonal cycle of SSTs proves an even more difficult challenge (Mechoso et al. 1995). Most models still have problems in accurately simulating the seasonal evolution of the eastern equatorial Pacific, the region key for the onset of ENSO. Many of the CGCMs simulate a significantly weaker annual cycle than is observed, while others simply fail to produce an annual cycle altogether. Some CGCMs produce too strong of an annual cycle, some a semiannual cycle, and others

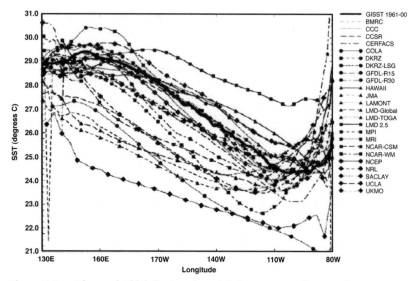

Figure 7.9. Observed (thick line) and modeled mean annual sea surface temperature (SST) in the equatorial (2°N–2°S) tropical Pacific, from Latif et al. (2001). Although the IPCC (2001) reports that coupled models provide credible simulations of mean annual equatorial Pacific climate, the range of modeled SSTs is more than 6°C.

phase shifts and displacements of the seasonal evolution of SSTs (Latif et al. 2001). A few models do simulate some aspects of the annual cycle realistically, and those tend to be ones with a high longitudinal resolution.

Teleconnections related to ENSO and resulting climate impacts occur at interannual time scales. Similar problems, however, are associated with modeling the interannual variability of the tropical Pacific. Again, those CGCMs with high spatial resolution (at least 0.5° latitude) provide reasonable interannual variability, but the remaining coarser resolution models cannot even simulate *past* ENSO variability. Some of these models produce a relatively regular ENSO frequency, while others produce a highly irregular interannual ENSO cycle. Other problems encountered are unrealistic spatial SST variability such as too strong/weak variability in some parts of the equatorial Pacific, no interannual variability at all, or too strong interannual variability (Latif et al. 2001). In addition, the level of high-frequency variability or "noise" is also very different among various CGCMs; of course, GCMs in general do not accurately simulate the statistics of weather noise (Slingo et al. 1996). Therefore, distinguishing between external variability (signal) and internal variability (noise) is very difficult for GCMs.

Overall, it is apparent that coupled atmosphere–ocean GCMs simply do not have the adequate longitudinal and latitudinal resolution required to accurately locate the ocean upwelling and strong gradients occurring before, during, and after El Niño events. Thus, while recent GCMs can be said to simulate ENSO-*like* fluctuations that capture some of the main features (Knutson and Manabe 1997), the simulated ENSO in fact tends to be weaker than what has been observed. The problem thus lies in the CGCMs' inability to generate both the climatology and the climate variability—both the annual cycle and the ENSO cycle. Many experiments with specified SSTs have illustrated that improving the mean climate in GCMs also improves the models' performance (Shukla 1998). It is a paradox that coupled GCMs that produce the annual cycle do not produce ENSO, while GCMs that produce the ENSO cycle do not produce the equatorial annual cycle (Battisti and Sarachik 1995).

Modeling Major Modes of Atmospheric Variability

PACIFIC/NORTH AMERICA

As difficult as it is for GCMs to accurately model ENSO and its signature in the tropics alone, their ability to accurately simulate the effects of ENSO outside of the tropical Pacific region is even more limited. One of the main observed extratropical responses to ENSO in the Pacific/North America region is a PNA-like pattern of atmospheric circulation. However, there is a wide range of conflicting GCM simulations, where some results illustrate a strong PNA-like response, while yet others indicate only a very weak extratropical sensitivity to ENSO forcing at all, and some find patterns distinctly different from the classic PNA pattern.

For instance, Hannachi (2001) employs multidecadal GCM runs in an effort to study the atmospheric response to ENSO-type tropical Pacific forcing. The classic PNA pattern emerges as a leading mode of ENSO-forced variability, thus supporting the traditional notion of El Niño being linked to the PNA teleconnection pattern. The modeling analysis further indicates that the PNA pattern is a synchronous response, locked to ENSO, with the positive phase of the PNA pattern occurring during El Niño, and the negative phase during La Niña. The amplitude of the response is directly proportional to the amplitude of the SST anomaly in the east tropical Pacific (Hannachi 2001). It must be noted that such a linear response is entirely inconsistent with the nonlinear nature involved in the relationship between the SST forcing and the atmospheric response (Hoerling et al. 1997).

Kumar and Hoerling (1997) employed an AGCM to test the atmospheric response of tropical Pacific El Niño forcing on the midtropospheric pressure field of the PNA region. They find that the AGCM composite pattern for El Niño years is not robust, illustrated by significant amplitude differences in their modeled pressure response, with a weaker Aleutian Low response and a much weaker North American response, which is attributable to model error. In general, an insensitivity of the atmospheric response pattern to inter-El Niño SST forcing is determined (Kumar and Hoerling 1997), consistent with other GCM studies (e.g., Geisler et al. 1985). The insensitivity of the atmosphere in this modeling study leads to the conclusion that inter-El Niño climate variations are consistent with noise—the internal variability of the atmosphere is larger than the external forcing, the signal, and that the atmosphere is simply not sensitive to tropical forcing (Hoerling and Kumar 2002). Furthermore, it supports the notion that El Niño variability can only modify existing, preferred extratropical patterns of internal variability.

The exact opposite conclusion was reached in the investigation of Straus and Shukla (2002). Not only does this modeling study find that tropical forcing can indeed generate distinct patterns of atmospheric circulation, but it is found that ENSO does not force the PNA, while La Niña in fact may. In their GCM study they find a dominant pattern of atmospheric variability associated with tropical El Niño forcing that is quite distinct from internal modes of atmospheric variability such as the PNA teleconnection pattern (figure 7.10). While their El Niño-forced pattern describes a wave train like the PNA pattern, the centers of action are significantly different.

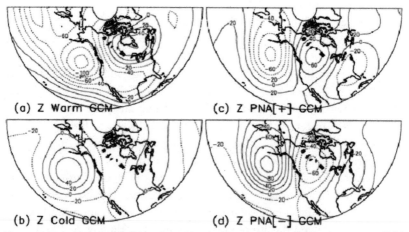

Figure 7.10.   The modeled Northern Hemisphere upper tropospheric pressure field as forced by ENSO (a and b), and by comparison the modeled PNA upper tropospheric pressure field (c and d). From Straus and Shukla (2002).

This results in a pattern of circulation that is much more north–south than the PNA pattern (Straus and Shukla 2002).

There is obvious and significant disagreement among GCM simulations about the exact nature of tropical Pacific SST forcing on the extratropical atmosphere in the PNA region. While the convective regions associated with ENSO variability in the tropical Pacific are key in the mechanism by which the tropics force the midlatitudes (see above), the models that actually capture ENSO precipitation anomalies in the central equatorial Pacific most realistically are also the models that fail to capture the extratropical responses to the SST forcing (AchutaRao and Sperber 2002). This shortcoming is in some cases manifested as a displacement of the location of the extratropical anomaly response, but in many cases the anomalies are of opposite sign, running completely counter to observations. This likely indicates a systematic error of the models' base state climatologies of the upper-level atmospheric flow, preventing propagation of the tropical signal to the extratropical Northern Hemisphere circulation (AchutaRao and Sperber 2002).

AO AND NAO

The ability of GCMs to accurately simulate changes in the NAO and AO are equally confounded as their ability to simulate the PNA. Some studies report that the "coupled climate models simulate the NAO quite well" (IPCC 2001), and show a close correspondence between observed and modeled indices describing the NAO—with correlations of 0.8; however, such close correspondence is only achieved after applying a seventy-three-month running mean filter on the data (Hoerling et al. 2001), a certainly dubious and likely fortuitous choice. Furthermore, this same study invokes global warming as the reason for the NAO's strong positive trend in both observed and filtered model time series, despite the fact that the observed NAO was largely negative during the late 1990s while CO_2 levels continued to increase.

Promisingly, thirteen out of seventeen models tested by Stephenson and Pavan (2003) had a leading pattern that captures main NAO features such as the northern dipole; yet the southern dipole is simulated less well and displaced too far eastward. Overall, NAO time series from coupled models are generally unrealistic due to monotonic trends and overly strong correlations between the tropics and extratropics, specifically with ENSO. That result led Stephenson and Pavan (2003), based on their comprehensive analysis of seventeen CGCMs, to reject the IPCC's claim that "coupled climate models simulate the NAO quite well." The core of the problem in modeling the NAO and its hemispheric counterpart, the AO, may lie in this oscillation's close ties between the troposphere and the stratosphere. The AO is observed to occur from the surface to the mid-stratosphere, with the changes in the stratosphere preceding AO changes in the troposphere, and propagating from the stratosphere to the troposphere. This obviously suggests that the AO could be forced from above, by stratospheric dynamics (Thompson and Wallace 1998).

Accordingly, Shindell et al. (2001) find that the observed wintertime surface pressure trend is reproduced only by models that are able to resolve stratospheric dynamics, that is, models with an upper boundary at 85 km, not the more common 30 km. Therefore, models that do not include sufficient stratospheric layers and thus cannot resolve the dynamics of that atmospheric layer uniformly fail to simulate the observed trend in the AO, even when this mode is present and dominates the wintertime variability in the model. While in the troposphere-only GCMs the typical response to increasing greenhouse gases is a greater surface warming at high latitudes, which does indeed lead to decreased surface pressure in the Arctic relative to midlatitudes, these pressure changes are too low in amplitude and are therefore only weakly correlated with the AO. In fact, the modeled trend of the AO time series is not statistically distinct from zero in many tropospheric GCMs despite the claim that GCMs simulate the NAO "quite well" (IPCC 2001). While some of these GCMs do contain (two) dynamical layers above the tropopause, two layers are insufficient to adequately resolve stratospheric dynamics (Shindell et al. 2001). The inability of these tropospheric GCMs to reproduce an AO trend comparable to the observed trend suggests that stratospheric dynamics must be accounted for, both for adequately modeling the atmospheric circulation response associated with the AO and NAO, and for detecting and predicting any future warming potentially associated with the AO/NAO. However, the current standard climate models do not have the sufficient vertical resolution to include and model stratospheric dynamics, therefore any model simulations regarding the AO (and NAO) are questionable.

All studies attempting to model the observed changes in the NAO and AO are based on increasing greenhouse gas concentrations—global warming—forcing those changes. However, evidence exists that demonstrates a significant solar influence, which is to say natural variability, on circulation features such as the AO. It is well known that variations in ultraviolet radiation caused by the solar cycle cause temperature and ozone changes in the stratosphere (e.g., Hood et al. 1993). However, a mechanism has been lacking that would allow stratospheric changes to propagate downward into the troposphere, thus influencing climate change. The AO provides this mechanism (Baldwin and Dunkerton 1999). Using greatly increased temporal resolution of forty years of daily data, and a spatial resolution that includes many atmospheric layers, Ruzmaikin and Feynman (2002) illustrate empirically that solar variations occurring on the timescale of years affect the AO. Specifically, the intensity of solar ultraviolet radiation changes the dominant modes of the AO. The important implications of these findings are that this strong solar influence on the AO means that greenhouse gas increases are not necessarily the cause of the long-term trends in the AO and NAO, and therefore any GCM simulations attempting to account for changes in the AO/NAO in this manner are inherently flawed in this basic assumption. Furthermore, and equally important, these findings provide further evidence that circulation-scale climate change is indeed significantly impacted by solar influences—natural variability.

THE STATE OF THE SCIENCE

For at least three decades, scientists have been attempting to numerically and dynamically simulate the behavior of the world's oceans and the overlying atmosphere, and their important interactions (e.g., Hoskins et al. 1977). But so far, their main accomplishment to date is recognizing that behavior's significant complexity. The extratropical response to large-scale forcing by the tropics is based on anomalous planetary wave propagation from the regions of tropical upper-air divergence as caused by ENSO. ENSO occurs largely, though by no means completely, due to air-sea coupling in the tropical equatorial Pacific and is acknowledged as the most predictable element of the climate system on seasonal to interannual time scales. Even so, there are profound gaps in our understanding of the predictability of the climate system as a whole, and ENSO in particular. Much of that lack of understanding derives from the large systematic errors in the models—especially the coupled models, which should be most realistic in that they model a changing ocean interacting with a changing atmosphere—and these errors affect our estimates of the limit of predictability.

It can be argued that some GCMs can indeed simulate the broad, large-scale patterns of tropical Pacific SST variability. In general, however, adequately simulating even the proper warm pool (El Niño) or cold tongue (neutral or La Niña) structure is still a challenge for models given their significant problems regarding the amplitude of the anomalies, and problems at the boundaries of the eastern and western equatorial Pacific. The eastern Pacific especially is very important in ENSO variability, given its role during the onset and generation of ENSO events. Most models still cannot accurately model the climatology of the tropical Pacific, despite claims that in general, CGCMs provide credible simulations of climate, at least down to subcontinental scales and over temporal scales from seasonal to decades (IPCC 2001).

The fact remains that neither the mean annual values, nor the seasonal cycle, nor the interannual variability of ENSO can be reliably simulated (Latif et al. 2001). Yet projections are made about future ENSO variability, indicating a shift toward a more El Niño-like regime that will result in future seasonal precipitation extremes that are likely to be more intense (IPCC 2001). While some models are capable of, to some degree, simulating some of these aspects of ENSO variability, no models are able to simulate them all. Partly, these shortcomings can be attributed to the spatial resolution in GCMs. Those models with a high meridional resolution tend to outperform the models with a coarser resolution, but that is by no means true of all models with high spatial resolution. Therefore, while high resolution is necessary, it is not sufficient in providing realistic simulations.

Given the models' shortcomings in reliably simulating the tropical variability in the Pacific, it follows that simulations of the extratropical responses to tropical forcing are also inadequate. Predictions for an extratropical response in the Northern Hemisphere atmospheric circulation in the PNA region occupy the whole range of possible scenarios. Some modeling studies suggest a strong PNA pattern resulting from El Niño forcing, other research suggests a very weak PNA response, while yet other research shows patterns entirely different from the traditional PNA teleconnection pattern. Ironically, those models that are actually able to simulate the convective response to SST forcing in the equatorial tropics are the ones that perform the worst in predicting the extratropical response. That paradox can be attributed to the incorrect simulation of the vertical profile of the tropical atmosphere's diabatic heating as well as systematic errors in the models' base state of upper-level atmospheric flow, which results in the incorrect propagation of the tropical signal to the extratropical atmosphere.

The inability of GCMs to model the atmosphere over the Atlantic Ocean (the NAO) or the hemispheric manifestation of that oscillation (the AO) may be another issue of inadequate spatial resolution—in this case, vertically. As there is a close dependence for this atmospheric mode between the stratosphere and the troposphere, it seems reasonable and essential that GCMs would need to properly simulate stratospheric variability, and its connection to the troposphere. To date, this issue has proven entirely too complex for GCMs. Some modeling studies are able to simulate trends similar to those of the NAO/AO, even without incorporating the stratosphere; however, those models all invoke greenhouse gas forcing and assume that the source of changing atmospheric circulation is solely anthropogenic, without regarding sources of natural variability. An opposing view is provided by, for instance, Ruzmaikin and Feynman (2002), who demonstrate that natural variability, specifically solar forcing, is a significant influencer of stratospheric winds and the AO. Therefore, further doubt is cast on GCM predictions based on only greenhouse gas forcing, without incorporating natural variability.

A further complicating factor for providing reliable GCM prognostications regarding external variability/forcing is that atmospheric circulation exhibits significant noise—that is, internal variability independent of any external forcing. A significant problem with current GCMs is that they cannot adequately simulate noise. That factor, combined with the aforementioned limitations, makes it hard to comprehend how assessments are made by institutions such as the IPCC about our present and future climate, assessments that often times lead to national and international policy changes. The IPCC (2001) considers coupled models to be suitable tools to provide useful projections of future climates. But the serious limitations of our modeling capabilities must be realized, and GCM simulations simply must not serve as the basis for such decisions. Our climate is changing, to be sure. But the source of those changes is at best uncertain, and until we can accurately model our current and future climate, our future should be based on past and current observed trends, not those simulated.

What, then, of the IPCC's statements about the connection of climate change with El Niño and other atmospheric circulations? In light of this paper's comprehensive analysis of those phenomena and the GCM's inability to model them, let us reconsider the IPCC's claims. Its *Third Assessment Report* concluded that "a majority of models show a mean El Niño-like response in the tropical Pacific, with the central and eastern equatorial Pacific sea-surface temperatures warming more than the western equatorial Pacific, with a corresponding mean eastward shift of precipitation." Yet the balance of the evidence contradicts this assertion. Consider also how the IPCC downplayed the level of scientific uncertainty in this crucial area. "Attempts to address this question using climate models have again shown conflicting results, varying from slight decreases in amplitude . . . to a small increase in amplitude." An understatement, to be sure. And as for the models' indicating

> that future seasonal precipitation extremes associated with a given ENSO event are likely to be more intense due to the warmer, more El Niño-like mean base state in a future climate . . . also in association with changes in the extratropical base state in a future warmer climate, the teleconnections may shift somewhat with an associated shift of precipitation and drought conditions in future ENSO events.

Based on the balance of evidence, there is insufficient support for this claim as well.

Without question, much more progress is necessary regarding our current understanding of climate and our abilities to model it. Before we can accurately understand the midlatitudes' response to tropical forcing, the tropical forcings themselves must be identified and understood. The atmospheric response must then be reliably determined, which involves various complicated processes such as how the internal modes of tropical variability are transferred into external variability modes in the extratropics. Furthermore, the tropical signal in the midlatitude atmosphere must be distinguished from the internal variability, and the various air-sea feedbacks must be better understood. Similarly, the internal and external modes of variability in the extratropical oceans must be distinguished. Only after we identify these factors and determine how they affect one another can we begin to produce accurate models. And only then should we rely on those models to shape policy. Until that time, climate variability will remain controversial and uncertain.

REFERENCES

AchutaRao, K., and K. R. Sperber. 2002. Simulation of the El Niño Southern Oscillation: Results from the Coupled Model Intercomparison Project. *Clim Dynamics* 19:191–209.

Alexander, M. A. 1992. Midlatitude Atmosphere–Ocean Interaction during El Niño. Part I: The North Pacific Ocean. *J Climate* 5:944–58.

Alexander, M. A., I. Bladé, M. Newman, J. R. Lanzante, N.-C. Lau, and J. D. Scott. 2002. The Atmospheric Bridge: The Influence of ENSO Teleconnections on Air–Sea Interaction over the Global Oceans. *J Climate* 15:2205–31.

Angell, J. K. 1981. Comparison of Variations in Atmospheric Quantities with Sea Surface Temperature Variations in the Equatorial Eastern Pacific. *Monthly Weather Review* 109:230–43.

Angell, J. K. 1999. Comparison of surface and tropospheric temperature trends estimated from a 63-station radiosonde network, 1958–1998. *Geophys Res Lett* 26: 2761–64.

Angell, J. K. 2000. Tropospheric temperature variations adjusted for El Niño, 1958–1998. *J Geophy Res* 105:11841–49.

Arkin, P. A., W. Y. Chen, and E. M. Rasmusson. 1980. Fluctuations in mid- and upper tropospheric flow associated with the Southern Oscillation. Proceedings of the Fifth Annual Climate Diagnostics Workshop, U.S. Department of Commerce, Washington, D.C.

Baldwin, M. P., and T. J. Dunkerton. 1999. Propagation of the Arctic Oscillation from the stratosphere to the troposphere. *J Geophys Res* 104:30937–46.

Barnett, T. P., L. Dümenil, U. Schlese, E. Roeckner, and M. Latif. 1989. The Effect of Eurasian Snow Cover on Regional and Global Climate Variations. *J Atmos Sci* 46:661–85.

Battisti, D. S., and E. S. Sarachik. 1995. Understanding and Predicting ENSO. *Rev Geophys* 33:1367–76.

Bjerknes, J. 1966. A possible response of the atmospheric Hadley circulation to equatorial anomalies of ocean temperature. *Tellus* 18:820–29.

Bjerknes, J. 1969. Atmospheric teleconnections from the equatorial Pacific. *Monthly Weather Review* 97:163–72.

Branstator, G. W. 1990. Low-frequency patterns induced by stationary waves. *J Atmos Sci* 47:629–48.

Caron J., and J. J. O'Brien. 1998. The generation of synthetic sea surface temperature data for the equatorial Pacific Ocean. *Monthly Weather Review* 126: 2809–21.

Clark, M. P., and M. C. Serreze. 2000. Effects of Variations in East Asian Snow Cover on Modulating Atmospheric Circulation over the North Pacific Ocean. *J Climate* 13:3700–3710.

Cohen, J., and D. Entekhabi. 1999. Eurasian snow cover variability and Northern Hemisphere climate predictability. *Geophys Res Lett* 26:345–48.

Davis, R. E. 1976. Predictability of sea surface temperature and sea level pressure anomalies over the North Pacific Ocean. *J Phys Oceanogr* 6:249–66.

Deser, C., and M. L. Blackmon. 1995. On the relationship between tropical and North Pacific sea surface temperature variations. *J Climate* 8:1677–80.

Emery, W. J., and K. Hamilton. 1985. Atmospheric forcing of interannual variability in the northeast Pacific Ocean: Connections with El Niño. *J Geophys Res* 90: 857–68.

Frauenfeld, O. W. 2003. Northern Hemisphere Circulation Variability and the Pacific Ocean. Ph.D. diss., University of Virginia.

Frauenfeld, O. W., and R. E. Davis. 2000. The Influence of El Niño–Southern Oscillation Events on the Northern Hemisphere 500 hPa Circumpolar Vortex. *Geophys Res Lett* 27:537–40.

Frauenfeld, O.W., and R. E. Davis. 2002. Midlatitude Circulation Patterns Associated with Decadal and Interannual Pacific Ocean Variability. *Geophys Res Lett* 29:2221, 10.1029/2002GL015743.

Frauenfeld, O. W., and R. E. Davis. 2003. Northern Hemisphere circumpolar vortex trends and climate change implications. *J Geophys Res* 108:4423, 10.1029/2002JD002958.

Geisler, J. E., M. L. Blackmon, G. T. Bates, and S. Muñoz. 1985. Sensitivity of January climate response to the magnitude and position of equatorial Pacific sea surface temperature anomalies. *J Atmos Sci* 42:1037–49.

Graham, N. E., T. P. Barnett, R. Wilde, M. Ponater, and S. Schubert. 1994. On the roles of tropical and midlatitude SSTs in forcing annual to interdecadal variability in the winter Northern Hemisphere circulation. *J Climate* 7:1416–42.

Gu, D., and S. G. H. Philander. 1997. Interdecadal climate fluctuations that depend on exchanges between the tropics and extratropics. *Science* 275:805–7.

Hanawa, K. T., Y. Yoshikawa, and T. Watanabe. 1989. Composite analyses of wintertime wind stress vector fields with respect to SST anomalies in the western North Pacific and the ENSO events. Part I: SST composite. *J Meteor Soc Japan* 67:385–400.

Hannachi, A. 2001. Toward a nonlinear classification of the atmospheric response to ENSO. *J Climate* 14:2138–49.

Hildebrandsson, H. H. 1897. Quelque recherches sur les entres d'action de l'atmosphere. *K Sven Vetenskaps akad Handl* 29:1–33.

Hoerling, M. P., and A. Kumar. 1997. Why do North American climate anomalies differ from one El Niño to another? *Geophys Res Lett* 24:1059–62.

Hoerling, M. P., and A. Kumar. 2002. Atmospheric response patterns associated with tropical forcing. *J Climate* 15:2184–2203.

Hoerling, M. P., A. Kumar, and M. Zhong. 1997. El Niño, La Niña, and the nonlinearity of their teleconnections. *J Climate* 10:1769–86.

Hoerling, M. P., J. W. Hurrell, and T. Xu. 2001. Tropical origins for Recent North Atlantic climate change. *Science* 292:90–92.

Hood, L. L., J. L. Jiricovich, and J. P. McCormack. 1993. Quasi-decadal variability of the stratospheric influence on long-term solar ultraviolet variations. *J Atmos Sci* 50:3949–58.

Horel, J. D., and J. M. Wallace. 1981. Planetary-scale atmospheric phenomena associated with the Southern Oscillation. *Monthly Weather Review* 109:813–29.

Hoskins, B. J., and D. J. Karoly. 1981. The steady linear response of a spherical atmosphere to thermal and orographic forcing. *J Atmos Sci* 38:1179–96.

Hoskins, B. J., and P. D. Sardeshmukh. 1987. A diagnostic study of the dynamics of the northern hemisphere winter of 1985–86. *Quart J Roy Meteor Soc* 113:759–78.

Hoskins, B. J., A. J. Simmons, and D. G. Andrews. 1977. Energy dispersion in a barotropic atmosphere. *Quart J Roy Meteorol Soc* 103:553–68.

Hurrell, J. W. 1996. Influence of variations in the extratropical wintertime tele-connections on Northern Hemisphere temperature. *Geophys Res Lett* 23:665–68.

IPCC. 2001. *Climate Change 2001: The Scientific Basis.* Contribution of Working Group I to the Third Assessment Report of the IPCC. Cambridge: Cambridge University Press.

Jacobs, G. A., H. E. Hurlbert, J. C. Kindle, E. J. Metzger, J. L. Mitchell, W. J. Teague, and A. J. Wallcraft. 1994. Decade-scale trans-Pacific propagation and warming effects of an El Niño anomaly. *Nature* 370:360–63.

Knutson, T. R., and S. Manabe. 1997. Simulated ENSO in a global coupled ocean-atmosphere model: Multidecadal amplitude modulation and CO_2 sensitivity. *Geophys Res Lett* 21:2295–98.

Kumar, A., and M. P. Hoerling. 1995. Prospects and limitations of seasonal atmos-pheric GCM predictions. *Bull Amer Meteor Soc* 76:335–45.

Kumar, A., and M. P. Hoerling. 1997. Interpretation and implications of observed inter-El Niño variability. *J Climate* 10:83–91.

Kumar, A., W. Wang, M. P. Hoerling, A. Leetmaa, and M. Ji. 2001. The sustained North American warming of 1997 and 1998. *J Climate* 14:345–53.

Lamb, P., and R. Peppler. 1991. West Africa. In *Teleconnections Linking Worldwide Climate Anomalies*, 121–89. Edited by M. H. Glanz, R. W. Katz, and N. Nicholls. Cambridge: Cambridge University Press.

Latif, M., K. Sperber, J. Arblaster, P. Braconnot, D. Chen, A. Colman, U. Cubasch, et al. 2001. ENSIP: the El Niño Simulation Intercomparison Project. *Clim Dynamics* 18:255–76.

Lau, N.-C. 1985. Modeling the seasonal dependence of the atmospheric response to observed El Niños in 1962–76. *Monthly Weather Review* 113:1970–96.

Lau, N.-C., and M. J. Nath. 1996. The role of the "atmospheric bridge" in link-ing tropical Pacific ENSO events to extratropical SST anomalies. *J Climate* 9: 2036–57.

Lean, J., and D. Rind. 1998. Climate forcing by changing solar radiation. *J Climate* 11:3069–94.

Levitus S., J. I. Antonov, T. P. Boyer, and C. Stephens. 2000. Warming of the world ocean. *Science* 287:2225–29.

Lockyer, N., and W. J. S. Lockyer. 1902. On the similarity of the short-period pres-sure variation over large areas. *Proc R Soc London* 71:134–35.

Lockyer, N., and W. J. S. Lockyer. 1904. The behavior of the short-period atmos-pheric pressure variation over the earth's surface. *Proc R Soc London* 73:457–70.

Madden, R. A. 1976. Estimates of natural variability of time-averaged sea level pressure. *Monthly Weather Review* 104:942–52.

McPhaden, M. J. 1999. The child prodigy of 1997–1998. *Nature* 398:559–62.

Mechoso, C. R., A. W. Robertson, N. Barth, M. K. Davey, P. Delecluse, P. R. Gent, S. Ineson, et al. 1995. The seasonal cycle over the tropical Pacific in cou-pled ocean-atmosphere general circulation models. *Monthly Weather Review* 123: 2825–38.

Michaels, P. J., P. C. Knappenberger, R. C. Balling, and R. E. Davis. 2000. Observed warming in cold anticyclones. *Clim Res* 14:1–6.

Miller, A. J., D. R. Cayan, T. P. Barnett, N. E. Graham, and J. M. Oberhuber. 1994. The 1976–77 climate shift of the Pacific Ocean. *Oceanography* 7:21–26.

Minobe, S. 1997. A 50–70-year climatic oscillation over the North Pacific and North America. *Geophys Res Lett* 24:683–86.

Minobe, S. 1999. Resonance in bidecadal and pentadecadal climate oscillations over the North Pacific: Role in climatic regime shifts. *Geophys Res Lett* 26:855–58.

Mo, K. C., and R. E. Livezey. 1986. Tropical–extratropical geopotential height teleconnections during the Northern Hemisphere winter. *Monthly Weather Review* 114:2488–515.

Namias, J. 1976. Some statistical and synoptic characteristics associated with El Niño. *J Physical Oceanogr* 6:130–38.

Niebauer, H. J. 1984. On the effect of El Niño events in Alaskan waters. *Bull Amer Meteor Soc* 65:472–473.

Niebauer, H. J. 1988. Effects of El Niño–Southern Oscillation and North Pacific weather patterns on interannual variability in the southern Bering Sea. *J Geophys Res* 93:5051–68.

Nitta, T., and S. Yamada. 1989. Recent warming of tropical sea surface temperature and its relationship to the Northern Hemisphere circulation. *J Meteor Soc Japan* 67:187–93.

Palmer, T. N., and J. A. Owen. 1986. A possible relationship between some "severe" winters over North America and enhanced convective activity over the tropical West Pacific. *Monthly Weather Review* 114:648–51.

Peng, P., A. Kumar, A. G. Barnston, and L. Goddard. 2000. Simulation skills of the SST-forced climate variability of the NCEP-MRF9 and the Scripps-MPI ECHAM3 models. *J Climate* 13:3657–79.

Philander, S. G. 1990. *El Niño, La Niña, and the Southern Oscillation*. San Diego: Academic.

Quinn, W. H., and V. T. Neal. 1984. Recent climate change and the 1982–83 El Niño, 148–54. Presented at the Eighth Annual Climate Diagnostic Workshop, Downsville, Ont.

Quinn, W. H., and V. T. Neal. 1985. Recent long-term climate change over the eastern tropical and subtropical Pacific and its ramifications, 101–9. Presented at the Ninth Annual Climate Diagnostic Workshop, Corvallis, Ore.

Quiroz, R. S. 1977. Tropospheric-stratospheric polar vortex breakdown of January 1977. *Geophys Res Lett* 4:151–54.

Rasmusson, E. M. 1991. Observational aspects of ENSO cycle teleconnections. In *Teleconnections Linking Worldwide Climate Anomalies*, 309–44. Edited by M. H. Glanz, R. W. Katz, and N. Nicholls. Cambridge: Cambridge University Press.

Reynolds, R. W., and E. M. Rasmusson. 1983. The North Pacific sea surface temperature associated with El Niño events, 298–310. Presented to the Seventh Climate Diagnostics Workshop, Boulder, Colo.

Ruzmaikin, A., and J. Feynman. 2002. Solar influence on a major mode of atmospheric variability. *J Geophys Res* 107:4209, doi:10.1029/2001JD001239.

Shindell, D. T., G. A. Schmidt, R. L. Miller, and D. Rind. 2001. Northern Hemisphere winter climate response to greenhouse gas, ozone, solar, and volcanic forcing. *J Geophys Res* 106:7193–210.

Shukla, J. 1998. Predictability in the Midst of Chaos: A Scientific Basis for Climate Forecasting. *Science* 282:728–31.

Shukla, J., and J. M. Wallace. 1983. Numerical simulation of the atmospheric response to equatorial Pacific sea surface temperature anomalies. *J Atmos Sci* 40: 1613–30.

Shukla, J. S., J. Anderson, D. Baumhefner, C. Brankovic, Y. Chang, E. Kalnay, L. Marx, et al. 2000. Dynamical seasonal prediction. *Bull Amer Meteor Soc* 81: 2593–606.

Simmons, A. J. 1982. The forcing of stationary wave motions by tropical diabatic forcing. *Quart J Roy Meteor Soc* 108:503–34.

Simmons, A. J., J. M. Wallace, and G. Branstator. 1983. Barotropic wave propagation and instability, and atmospheric teleconnection patterns. *J Atmos Sci* 40: 1363–92.

Slingo, J. M., J. S. Boyle, J.-P. Ceron, M. Dix, B. Dugas, W. Ebisuzaki, J. Fyfe, et al. 1996. Intraseasonal oscillations in 15 atmospheric general circulation models: Results from an AMIP diagnostic subproject. *Clim Dynamics* 12:325–57.

Stephenson, D. B., and V. Pavan. 2003. The North Atlantic Oscillation in coupled climate models: a CMIP1 evaluation. *Clim Dynamics* 20:381–99.

Straus, D. M., and J. Shukla. 2002. Does ENSO force the PNA? *J Climate* 15: 2340–58.

Thompson, D. W. J., and J. M. Wallace. 1998. The Arctic Oscillation signature in the wintertime geopotential height and temperature fields. *Geophys Res Lett* 25: 1297–1300.

Thompson, D. W. J., and J. M. Wallace. 2001. Annular modes in the Extratropical Circulation. Part II: Trends. *J Climate* 13:1018–36.

Trenberth, K. E. 1993. The different flavors of El Niño, 50–53. Presented to the Eighteenth Annual Climate Diagnostics Workshop, Boulder, Colo.

Trenberth, K. E. 1997. Short-term climate variations: Recent accomplishments and issues for future progress. *Bull Amer Meteor Soc* 78:1081–96.

Trenberth, K. E., G. W. Branstator, D. Karoly, A. Kumar, N.-C. Lau, and C. Ropelewski. 1998. Progress during TOGA in understanding and modeling global teleconnections associated with tropical sea surface temperatures. *J Geophys Res* 103:14291–324.

Trenberth, K. E., and T. J. Hoar. 1996. The 1990–1995 El Niño–Southern Oscillation event: Longest on record. *Geophys Res Lett* 23:57–60.

Wallace, J. M., and D. S. Gutzler. 1981. Teleconnections in the geopotential height field during the Northern Hemisphere winter. *Monthly Weather Review* 109: 784–812.

Wallace, J. M., and Q.-R. Jiang. 1987. On the observed structure of the interannual variability of the ocean/atmosphere climate system. In *Atmospheric and Oceanic Variability*, 17–43. Edited by H. Cattle. Bracknell, Berkshire, UK: Royal Meteorological Society.

Wallace, J. M., and S. Vogel. 1994. *Reports to the nation on our changing planet: El Niño and climate prediction*. NA27GPO232-01. Boulder, Colo.: National Oceanic and Atmospheric Administration.

Weare, B. C., A. Navato, and R. E. Newell. 1976. Empirical orthogonal analysis of Pacific Ocean sea surface temperatures. *J Phys Oceanogr* 6:671–78.

Webster, P. J. 1981. Mechanisms determining the atmospheric response to sea surface temperature anomalies. *J Atmos Sci* 38:554–71.

Yin, Z. 1994. Moisture condition in the South-Eastern USA and teleconnection patterns. *Int. J Climatol* 14:947–67.

Zhang, Y., J. M. Wallace, and D. S. Battisti. 1997. ENSO-like interdecadal variability: 1900–93. *J Climate* 10:1004–20.

Zhang, Y., J. M. Wallace, and N. Iwasaka. 1996. Is climate variability over the North Pacific a linear response to ENSO? *J Climate* 9:1468–78.

8

CLIMATE CHANGE
AND HUMAN HEALTH

Robert E. Davis

Global warming is one of the most important scientific issues of our time. The increasing atmospheric concentration of greenhouse gases (carbon dioxide, methane, nitrogen oxides, ozone, etc.) has been implicated in the temperature increases observed in the latter part of the twentieth century. Many of the greenhouse gas increases can be directly related to human activities, such as fossil fuel combustion, certain agricultural practices, land-use change, cattle production, and so on. A substantial number of climatologists believe the warmth of the late twentieth century is at least partially related to anthropogenic factors, although convincing proof remains a matter of debate within the climate community.

Whatever their cause, changes in greenhouse gas levels may impact the entire global climate system, potentially leading to changes in precipitation, atmospheric humidity, cloud cover, wind flow patterns, chemical reaction rates, sea level, and ocean circulation. Indeed, some of those changes already have been observed. Although no heat wave, for example, can be linked to anthropogenic climate change, changes in the number or intensity of heat waves over many decades is viewed by some as evidence of climate changes related to greenhouse gases. The critical question, however, is how climate change ultimately will impact society.

Of all the projected impacts of climate change, clearly the most important are those that could influence human lives and livelihood. Whereas some researchers associate climate change with increased human death and suffering, other scientists counter that human societies will thrive in the future climate. For example, the Intergovernmental Panel on Climate Change (IPCC) Working Group II (McCarthy et al. 2001), identifies a wide array

183

of possible factors that could have impacts on human societies, including floods, droughts, crop yields, availability of potable water, energy demand, timber supplies, mortality, tourism shifts, tropical cyclones, midlatitude cyclones, El Niño impacts, and infectious disease transmission, among others. It is a certainty that, given some degree of climate change, some individuals, groups, and societies will thrive while others will fail. That principle of adaptive response to environmental stressors is a fundamental component of evolutionary change. "More people are projected to be harmed than benefited by climate change," the IPCC predicts, "even for global mean temperature increases of less than a few °C." A dire prospect, perhaps, but then they assign only "low confidence" to that conclusion (5 percent to 33 percent likelihood), a lack of certainty that highlights the inherent complexities in modeling human responses (particularly behavioral responses) to environmental stressors and the difficulty of isolating a unique component of a human response to climate given that any response is likely to be driven by many different factors.

In this report, I explore a selected sample of human health impacts commonly expected to arise from climate change associated with increasing concentrations of atmospheric greenhouse gases. I review both the prevailing and alternative viewpoints to each issue, paying particular attention to the main topics highlighted in the Human Health chapter of IPCC Working Group II (McCarthy et al. 2001). It will very quickly become apparent that adaptations, whether examined from the perspective of organisms or human societies, will be key in determining the ultimate impacts of climate change. In several cases, adaptations have already taken place that have effectively mitigated against negative impacts from global warming.

HUMAN MORTALITY

Heat

Increasing death rates are the most important projected consequence of greenhouse warming. There is a well-established relationship between both heat waves and isolated hot events and higher mortality rates (Oechsli and Buechley 1970; Bridger et al. 1976; Kalkstein and Davis 1989; Katsouyanni et al. 1993; Kunst et al. 1993; Kalkstein and Greene 1997; Nakai et al. 1999; Davis et al. 2002, 2003a; Laschewski and Jendritzky 2002). When heat is coupled with high humidity, the body's ability to thermoregulate, or regulate its inner temperature, is compromised, resulting in excess physical stress. That effect is most notable among the elderly and the very young (Henschel et al. 1969; Oeschli and Buechley 1970; Lye and Kamal 1977; Applegate et al. 1981; Jones et al. 1982; Greenberg et al. 1983; Kunst et al. 1993; Kilbourne 1997). There is evidence that exposure to several consecutive hot and therefore stressful days results in higher mortality than might occur after an isolated but extremely hot and humid day, although little research has been performed on this topic. Minimum temperatures may be more closely related to mortality than maximum temperatures, as exposure to consecutive warm evenings allows the body little recovery time from the thermal stresses experienced during the day. Thus, morning dew point temperature (a measure of the amount of moisture in the air) is positively correlated with high mortality, particularly when dew point temperatures are high for several consecutive days (Kalkstein and Davis 1989; Kalkstein 1991; Smoyer et al. 2000).

Some causes of death appear to be more closely linked to weather parameters than others. Most of the cause-specific research has focused on diseases of the circulatory and respiratory systems (Donaldson and Keatinge 1997; Donaldson et al. 1998; Eng and Mercer 1998; Danet et al. 1999; Kloner et al. 1999; Lanksa and Hoffmann 1999; McGregor 1999, 2001; Pell and Cobbe 1999). Because of inconsistencies in the application of temporally and spatially consistent coding of mortality causes, "all-causes" (or total) mortality is most frequently used in weather and climate studies, and in some cases exhibits a stronger relationship to climate than the more "weather-related" mortality subcategories (Gover 1938; Schuman et al. 1964; Schuman 1972; Kalkstein and Davis 1989; Kunst et al. 1993; Kilbourne 1997; Davis et al. 2002, 2003a).

Given the well-established historical relationship between high heat and humidity and mortality, and based on model prognostications of higher temperatures arising from increasing greenhouse gas levels, mortality rates have also been projected to rise (Kalkstein and Greene 1997; Gaffen and Ross 1998; National Assessment Synthesis Team [NAST] 2000). Those mortality rates inherently account for background changes in the age demographics of the population and population changes, so they represent increases in the overall death rates. For example, in the U.S. National Assessment (NAST 2000), based on climate model projections of increasing temperatures, death rates are estimated to increase by 200 percent to 300 percent in some U.S. cities by 2050 relative to the historical baseline rate. Using an air mass–based approach, Kalkstein and Greene (1997) projected approximately a doubling of the summer mortality rate for the U.S. urban population by 2050, depending on the climate model chosen.

Observed and projected increases in humidity, coupled with higher temperatures, have been identified as a potential cause of excess heat mortality. Gaffen and Ross (1998), noting long-term increases in apparent temperature, or AT (an index that combines temperature and humidity and which serves as the basis of the Heat Index in the United States), in U.S. cities, suggested that a continuation of these trends "may pose a health problem, particularly as there are increasing numbers of elderly people, who are most vulnerable to heat-related sickness and mortality." Selected climate models project AT increases of more than 25°F in some regions by the end of this century (NAST 2000).

Because of the availability of high-quality daily data over a fairly long period of record, a significant portion of mortality research has been focused on the United States. High-quality studies also have been undertaken in other, primarily industrialized nations (e.g., Katsouyanni et al. 1993; Kunst et al. 1993; Nakai et al. 1999; Lashewski and Jendritzky 2002), but the lack of quality data in underdeveloped countries makes impact projection difficult for many regions. Furthermore, because of the need to examine mortality on a daily basis, the historical emphasis has been on urban mortality rates, where daily death totals are high enough to produce large and robust daily statistical samples. Higher urban temperatures arising from the heat-island effect, which is most pronounced in higher nighttime temperatures, coupled with higher humidity and lower air quality (see below), suggest that urban areas will be more highly impacted by climate change than rural regions (Oke 1987; Kilbourne 1997). But recent research indicates that rural areas have mortality rates that are comparable to urban regions and thus merit more careful study (Sheridan and Dolney 2003).

An inherent and fundamental problem in any studies that project future mortality is the assumption that people will respond to heat stress in the future in the same manner as they have in the past. But in a series of papers, Davis et al. (2002, 2003a, b) demonstrate that this assumption is invalid, at least in U.S. cities. Figure 8.1 is a plot of daily mortality anomalies (after age standardization to account for demographic changes, the daily death totals are subtracted from the monthly median because of the inherent seasonal cycle, see below) organized by three decades: the 1960s–1970s (10 total years), the 1980s (10 years), and the 1990s (9 years). The vertical line represents the Threshold Apparent Temperature (TAT), or the afternoon AT beyond which mortality increases on average by a statistically significant amount. In this example, the TAT for Philadelphia, established in the 1960s–1970s, is 32°C, and in that ten-year period, there is a significant increase in daily deaths when ATs exceed that value. But the response is less

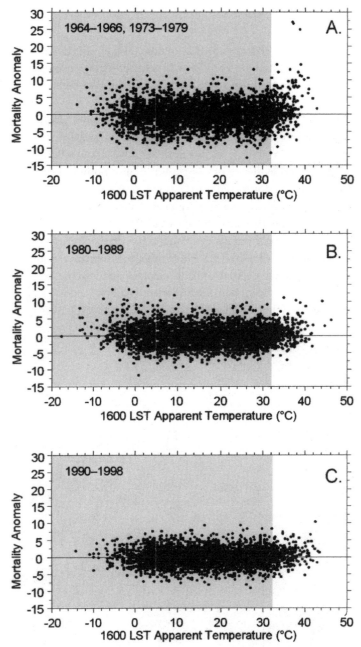

Figure 8.1. Daily mortality anomalies (daily deaths from all causes minus the monthly median) vs. 1600 Local Standard Time (LST) apparent temperature (AT) for Philadelphia, organized by "decade" (a: 1960–1970s, b: 1980s, c: 1990s). Areas with a white background show deaths that occur above the threshold apparent temperature, or the temperature above which mortality significantly increases. In this example, the threshold AT of 32°C was fixed in the 1960–1970s. Note that the mortality response to high ATs systematically declines over time, such that by the 1990s there is little evidence of excess mortality when ATs are high. (Source: Davis et al. 2002).

pronounced in the 1980s (figure 8.1, middle) and almost imperceptible in the 1990s (figure 8.1, bottom). Thus, in this particular example, the assumption of a temporally constant mortality response to high heat and humidity is incorrect.

Using this approach, Davis et al. (2003a) examined twenty-eight of the largest metropolitan areas in the United States and found significant mortality declines over time in twenty-two cities (figure 8.2). In eleven cities, there was no evidence of elevated mortality on hot and humid days in the 1990s. Overall for the cities examined, heat-related excess deaths declined from fifty-three (per million population per year) in the 1960s–1970s to twenty-five in the 1980s to fifteen in the 1990s.

The declining heat sensitivity of the U.S. urban populace to high ATs is attributable to numerous factors. Health care has improved, resulting in an overall reduction in the death rate that is incorporated into the heat response. Air-conditioning has become more pervasive, including a greater prevalence in northern cities, where it once was considered more of a luxury than a necessity. Studies in which the specific impact of air-conditioning on mortality

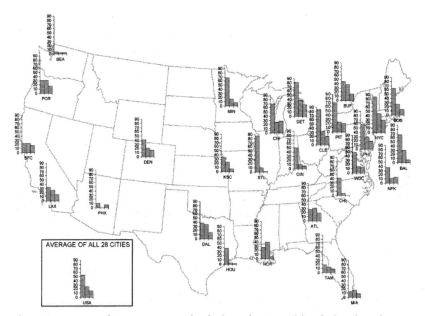

Figure 8.2.   Annual average excess deaths in major U.S. cities during days that exceed the threshold apparent temperature in each decade. The average of all twenty-eight cities is shown in the lower left. Populations have been age-standardized relative to a standard population to allow for direct comparisons over time and between cities. (Source: Davis et al. 2003a).

reduction was estimated vary significantly from an approximate 20 percent mortality reduction to almost 100 percent, or complete independence between mortality and heat (Kilbourne et al. 1982; Kalkstein 1993; Rogot et al. 1992; Semenza et al. 1996; Chan et al. 2001). Other factors, such as changes in architectural designs and materials, the dissemination of heat advisories and heat watch-warning systems, and the opening of air-conditioned shelters, have all produced some mortality benefit (Kalkstein et al. 1996; McGeehin and Mirabelli 2001). Human biophysical adaptation, or acclimatization, also plays a role in the human adaptive response. In total, these technological and adaptive practices have largely isolated the U.S. urban population from the effects of high heat and humidity on mortality from the 1960s through the 1990s.

A slightly different analytical approach is one in which the TAT is allowed to vary by decade. Given that background temperatures and humidity are generally increasing, we might propose alternate models of population adaptation. Davis et al. (2002, 2003a) used a TAT established in the early decades and applied it to subsequent time periods. But it is possible that people would adapt to increasing heat and humidity over time and yet for excess mortality to still exist (i.e., the TAT would actually change over time). Therefore, Davis et al. (2003b) used a decadally varying TAT and calculated the excess deaths arising from this partial adaptation model. For the United States as a whole (28 major cities), they found the number of excess deaths still declined significantly over time, from forty-one (per standard million population) in the 1960s–1970s, to seventeen in the 1980s, to eleven in the 1990s. While almost all cities had excess heat mortality in the first two decades, by the 1990s, the excess deaths were concentrated in cities in the northeastern United States and along the West Coast.

The spatial patterns of excess heat-related deaths are suggestive of a temporally varying adaptive response. By the 1980s, there was no evidence of excess mortality in southern cities where heat and/or humidity are common, such as Phoenix, Houston, Tampa, and Miami, where heat mortality had been a factor in the 1960s–1970s. By the 1990s, this full adaptation response had extended into more northern and interior cities, like Washington, D.C., Kansas City, and St. Louis. Again, air-conditioning penetration is likely a major factor in the mitigation of excess heat mortality in many U.S. cities.

A fundamental principle of population ecology is that the susceptible individuals within a population are most likely to be impacted by an environmental stressor. With respect to heat mortality, it is common that the mortality increase in response to a heat wave event is followed by several days of below-normal mortality, given that the remaining population is healthier and less susceptible. That effect is often called "mortality displacement" (Gover 1938; Schuman et al. 1964; Schuman 1972; Marmor 1975; Lyster 1976; Kalkstein 1993; Kilbourne 1997). For example, Philadelphia was influenced by a heat wave in early July 1966 that was associated with a pronounced mortality spike from July 6 through 8 (figure 8.3). But note that the days following the mortality peak in late July exhibited well below normal mortality. Thus, the net impact of the heat wave is actually the excess heat-related deaths minus the mortality displacement deaths. In this example, the mortality displacement was roughly 70 percent of the heat-related death total. So researchers who simply count the excess deaths on hot days are often overestimating the actual mortality. Although mortality displacement is a common occurrence, it is not found universally in all cities or for all heat waves (e.g., Laschewski and Jendritzky 2002). More research is needed on this topic to allow for the generation of better estimates of the actual, net impact of heat waves on mortality.

Cold

Globally, the warming experienced over the last half century has been predominantly concentrated in the cold half of the year (Balling et al.

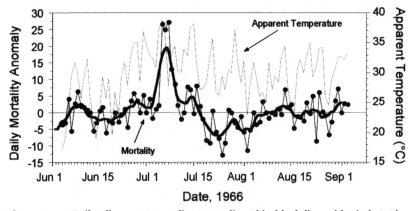

Figure 8.3.   Daily all-causes mortality anomalies (thin black line with circles) (departures from the monthly median) for Philadelphia in summer, 1966 and 1600 LST apparent temperatures (gray line). A seven-day smoother is applied to the mortality time series and is depicted as a thick black line. There is a notable mortality peak in early July related to the consistently high apparent temperatures, but there is also evidence of lower than normal mortality ("mortality displacement") in late July.

1998). Furthermore, Michaels et al. (2000) proposed that more warming occurs over high-latitude land regions where cold high-pressure systems (anticyclones) are frequently present. Theoretically, this component of the warming is likely related to greenhouse gas increases because of the lack of water vapor (the major greenhouse gas) in these air masses. Thus, a warming of the cold, winter high-pressure systems that commonly reside over East-Central Asia and northwestern North America may be a greenhouse gas response. Given that the southward movement of those air masses is responsible for most of the cold air outbreaks that occur in winter, we might project that these outbreaks will become milder in the future.

Weather-mortality relationships in winter are significantly more complicated than in summer. In nearly all cities examined globally, winter mortality is much higher than summer mortality, for reasons that are not well understood (Langford and Bentham 1995; Donaldson and Keatinge 1997; Eurowinter Group 1997; Lerchl 1998; Laschewski and Jendritzky 2002). Influenza and related respiratory diseases have a pronounced winter peak, and other causes of death also tend to be correlated with influenza epidemics (Simonsen et al. 2001). To date, no research group has been able to link the strength or duration of influenza epidemics or seasonal influenza deaths to weather or climate factors. People tend to spend more time in closed spaces in winter, thus increasing the opportunity for infectious disease to spread. Unlike summer, however, there is little relationship between daily weather and mortality in winter (Kunst et al. 1993; Kalkstein and Davis 1989). Aside from obvious cases of deaths related to automobile accidents during ice storms, heart attacks from snow shoveling, and so on, which add little to the total death counts, it is difficult to relate daily weather to daily mortality in winter in most studies. Some researchers have identified slight mortality increases arising from lower temperatures (Frost and Auliciems 1993) while a few others have found stronger relationships (Keatinge et al. 1989; Kunst et al. 1993). However, the results are not consistent and the winter relationships are generally weaker (Laschewski and Jendritzky 2002).

The nature of the observed warming (winter-dominant) coupled with the seasonality of mortality (high in winter, low in summer) has led some researchers to suggest that the lives lost in summer to excess heat might be more than mitigated by the lives saved during warmer winters (Langford and Bentham 1995; Martens et al. 1997; Guest et al. 1999; Donaldson et al. 2001). But the strength of that response is dependent on the extent to which weather is linked to winter mortality. While some researchers think the relationship is robust, others propose that the weak winter relationship implies that summer heat will continue to exert a greater impact relative to the few additional lives saved in the cold season (Kalkstein and Greene 1997).

In a current study on the seasonality of deaths for twenty-eight major U.S. cities, some weak relationships between monthly temperatures and monthly mortality have been identified (Davis et al. 2004). By examining monthly data, the total impact of heat or cold, including any mortality displacement effects, is captured. After removing the technological-adaptation trend from the data, Davis et al. found that the net impact of the observed warming from 1964 to 1998 resulted in 2.9 extra deaths (per standard million) per city. Thus, the observed warming pattern has resulted in excess summer deaths slightly exceeding the winter mortality reduction. But relative to the mortality baseline, the magnitude of the climate change effect is extremely small.

The intra-annual pattern of temperature change will be critical in determining the net annual mortality impact. Figure 8.4 shows the monthly temperature-mortality relationship averaged for U.S. cities. Higher temperatures are linked to higher mortality from June to September and lower mortality from October through May. Thus, 1°C of warming (with 75 per-

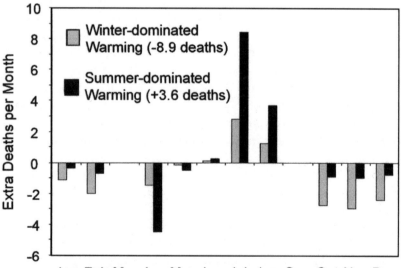

Figure 8.4.   **Number of excess or reduced monthly deaths under two different climate change scenarios resulting in a 1°C rise in temperature. In the winter-dominated case, 75 percent of the warming occurs in the six coldest months of the year, in the summer-dominated case, 75 percent of the warming occurs in the six warmest months of the year. (Source: Davis et al. 2003c)**

cent of the temperature increase concentrated in the warm half of the year) would generate 3.6 excess deaths (per standard million), but a 75 percent winter-dominated warming would result in 8.9 fewer deaths.

For the United States, at present, weather and climate have little net impact on mortality rates. The population's relative affluence has allowed for sufficient adaptations and technologies to be put into place that almost completely isolate the populace from adverse mortality impacts of weather and climate. In some cases, those adaptations have been implemented over relatively short periods of time. For example, Palecki et al. (2001) compared the 1995 and 1999 heat waves in Chicago and St. Louis. In both cases, but particularly in Chicago, high ATs in 1995 over a prolonged period had a major mortality impact (Semenza et al. 1996; Palecki et al. 2001). By some estimates, however, the 1999 heat event, though climatologically comparable, was associated with significantly lower mortality rates. According to Palecki et al. hundreds of lives were saved in Chicago in 1999 by the actions of city workers and because of media announcements. But in 1995, the death rate in Chicago (8.8 per 100,000) far exceeded that of St. Louis (1.1 per 100,000). The high death total in 1995 in Chicago can partially be attributed to a major electrical outage during the time of peak ATs, further emphasizing the importance of air-conditioning in mitigating heat mortality.

In the United States, the net impact of climate change on human mortality is currently very small and likely will remain so in the twenty-first century. Technological and adaptive responses of society have largely resulted in a populace that is immune from the impacts of extreme heat and cold. The lack of high-quality, long-term records for many locations makes it difficult to speculate on mortality impacts in many other nations. In Germany, Canada, and Great Britain, most research has emphasized cold season mortality and suggests that warmer winters will result in net mortality reductions (Langford and Bentham 1995; Guest et al. 1999; Donaldson et al. 2001; Laschewski and Jendritzky 2002). Little information is available for the third world. In general, however, because they are more directly exposed to the outdoors, the poor are more likely to suffer from thermal stresses and will probably exhibit higher death rates during weather extremes. Recall, however, the role of biophysical adaptation, which cannot be overlooked; then again, it is not well understood, so any estimates of future mortality based solely on current rates will probably be an overestimate.

Because air-conditioning requires electricity and produces heat, extensive use of air-conditioning raises the local environmental temperature to some degree and typically depends on fossil fuel sources for power. Clearly the net impact of air-conditioning on mortality is beneficial, but extensive air-conditioning usage, particularly in dense urban areas, requires reliable energy and results in some net heat release to the environment. Efforts to develop cost-efficient methods of bringing air-conditioning, in various forms, to Third World nations could result in very substantial future mortality reductions within this susceptible segment of the population.

AIR POLLUTION

Several trace constituents of the atmosphere, through their interactive relationship with factors that might change with increasing greenhouse gases, could grow in concentration and thereby impact human health. Most air quality and health research has focused on ozone and total suspended particulates (TSP), although some work has also been done on other pollutants like various sulfur oxides, nitrous oxides, and aeroallergens. Of course, carbon dioxide is not an atmospheric pollutant, but rather a naturally occurring and necessary component of the atmospheric constituent mix.

There remains significant debate regarding the extent to which pollutants impact human mortality, although there is less disagreement regarding pollutants and morbidity. Earlier work suggested that ozone, SO_2, and to some extent NO_2 were linked to higher mortality rates; however, the emphasis has more recently shifted to particulate matter, mostly total suspended particulate (TSP) levels and PM_{10} (suspended particulate matter greater than 10 microns in diameter).

For example, Pope et al. (1991) and Pope and Dockery (1992) found significant declines in lung function in adults and children linked to PM_{10}, even at levels below the national standards set by the U.S. Environmental Protection Agency. In highly publicized research, Dockery et al. (1993) linked air pollution rates in six U.S. cities to mortality rates from lung cancer and cardiopulmonary disease, noting that PM_{10} was the pollutant exhibiting the strongest relationship to morbidity and mortality. Dockery et al. (1989) also found stronger linkages in childhood illness to particulate levels as compared with SO_2 and NO_2. But Moolgavkar et al. (1995) proposed that particulates are often incorrectly singled out as causative because they are highly correlated with other constituent pollutants. After standardizing for temperature, season, and pollutant intercorrelations, Moolgavkar et al. (1995) found ozone (in summer) and SO_2 (in the other seasons) to be related to higher mortality levels in Philadelphia.

More recently, Samet et al. (2000) identified mortality increases related to higher PM_{10} levels in a twenty-city study—despite generally declining

PM_{10} levels (Ware 2000). Interestingly, of the five pollutants examined, only PM_{10} was significantly related to mortality; even summer ozone levels were not statistically significant.

The major complication in the pollution–mortality arena is the extent to which weather is actually a confounding factor in the analysis. Pollutant levels are concentrated when the height of the atmospheric mixed layer (the volume of air into which the pollutants are released) is reduced; this typically occurs in conjunction with atmospheric temperature inversions (temperatures increasing with height). Inversions substantially reduce the mixed-layer height, and the stable atmospheric layer discourages the dilution of pollutant concentrations. In general, however, inversion situations are *not* the warmest days. Thus, there is debate as to whether heat or pollution is the more critical factor in summer mortality rates.

In a study of daily deaths in Athens, Katsouyanni et al. (1993) found that air pollution was not related to mortality in general, but that mortality increased with high SO_2 levels when temperatures were also high. They found no relationship with ozone. Ibald-Mulli et al. (2001) also concluded that weather confounds understanding of the relationships between pollutants and health effects. Conversely, Samet et al. (1998), using daily data from Philadelphia, confirmed their earlier work and claimed that weather was not, in fact, a confounding factor. But in a similar, comparative study, Pope and Kalkstein (1996) came to the opposite conclusion.

As an example of the potential confounding problem, Smoyer et al. (2000) first identified the air mass most responsible for high summer mortality in Philadelphia (figure 8.5a). In her analysis, a hot and humid air mass exhibited significantly higher mortality than any of the other eight air masses. But when levels of ozone and TSP within this offensive air mass were examined, mortality was not systematically different for pollution levels in the highest or lowest quintiles (figure 8.5b).

Given the debate regarding (1) whether it is pollutants or heat and humidity that are directly responsible for higher mortality rates and (2) if pollutants are involved, which ones are most important, the prognostication of future climate change impacts on pollution and health is difficult at best. High temperatures tend to increase the rates of certain chemical reactions involved in the formation of certain pollutants. The ozone formation reaction requires sunlight, so future cloud cover will be a factor. A key consideration will be any changes in the frequency or intensity of inversions. In general, if observations are correct, global surface temperatures are increasing faster than temperatures in the free atmosphere (Hegerl and Wallace 2002a, b). The result is a net destabilization of the

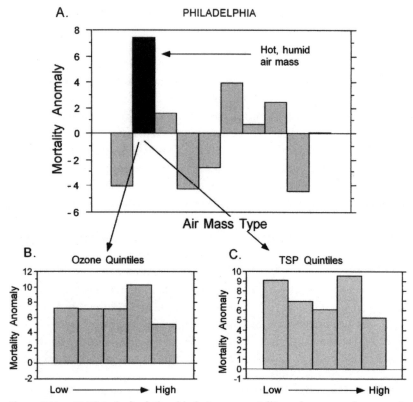

Figure 8.5.   A) Historical relationship between mortality and summer air masses in Philadelphia. The black bar indicates the air mass (noted for high temperatures and humidity) with the strongest mortality response. B, C) Mortality anomaly for days within the hot, humid air mass stratified into ozone and total suspended particulates (TSP) quintiles. There is no evidence of an air pollution/mortality relationship, suggesting that weather is a more important factor than air quality in this example (data from Smoyer et al. 2000).

atmosphere that would result in the dilution of pollutant levels through a greater atmospheric volume. Of course, some regions may experience changes that could result in higher inversion frequencies. Superimposed on these atmospheric changes are long-term trends in pollutant levels. Most industrialized nations have taken significant measures to improve air quality, but third world nations with nascent industrialization are less likely to be concerned with future air quality impacts at the cost of economic advancement.

Asthma occurrence is linked to weather and the biosphere in a complicated manner. Pollen production is clearly linked to weather and results in seasonal allergy problems for a growing sector of the population. At different times of year, different weather situations are implicated in asthma attacks. Linkages have been found with weather following cold front passages (Goldstein 1980), cool dry air (Carey and Cordon 1986), and thunderstorms (Hajat et al. 1997), whereas other studies find little weather relationship to asthma (Epton et al. 1997). Air pollution does not appear to exacerbate asthma occurrence or severity (Dawson et al. 1983; Steib et al. 1996; Anderson et al. 1998; Hajat et al. 1999). Indoor allergens are increasingly being implicated in asthma occurrence, particularly cockroach allergens (Platts-Mills and Carter 1997; Rosenstreich et al. 1997).

The complexity of the relationships between aeroallergens, weather and climate, and the confounding influence of socioeconomic and local/regional factors, makes the prediction of climate change impacts on health practically impossible given the current level of understanding.

In summary, the lack of simple, direct relationships between pollutants and mortality makes the prediction of future impacts extremely difficult. Retrospective analyses tend to find weak relationships between air quality and mortality. Changes in future weather patterns will influence those relationships, as will technological changes that will result in cleaner-burning fuels and improved medical care. Efforts for nations undergoing nascent industrialization to utilize modern technologies should be encouraged.

VECTOR-BORNE DISEASE TRANSMISSION

The transmission of a certain class of diseases requires an intermediary organism (or "vector") to produce human infection. In many cases, insects (often mosquitoes, ticks, or fleas) serve in the role of vectors. To some extent, the species range, abundance, activity rate, breeding rate, and so on are dependent on climatic factors such as temperature, precipitation, length of active season, and humidity.

Malaria is the most commonly cited vector-borne disease, and because of its prevalence in many of the earth's tropical and subtropical climates, in highly populated areas, it has garnered significant attention. Other important vector-related diseases are dengue, leishmaniasis, yellow fever, and more recently West Nile virus.

How vector-borne disease transmission is related to weather and climate factors, and how projected climate changes might impact vector-borne disease transmission, are perhaps the most well-studied topics in the suite of climate change impacts. But as with all the other topics in this chapter, there are numerous factors unrelated to climate that significantly influence the process. Here, we review how weather and climate influence vector-borne diseases, and also evaluate potential confounding factors that could bias impacts that are often assigned to climatic changes.

Malaria

The risk of malaria transmission from a single mosquito to a human can be estimated by multiplying the mosquito biting rate by the transmission rate (not all bites transmit the virus) by the mosquito density per human modified by the mosquito mortality rate by the incubation period (the time from when the pathogen begins to develop in the mosquito until the mosquito becomes infective). Most of the variables in this equation are temperature-sensitive (Rogers and Packer 1993; Reiter 2001). But it is incorrect to assume that the net impact of a temperature increase will be increased infection. For example, mosquitoes assume substantial risk by biting, and individual mosquitoes (or populations) with high biting rates also have high mortality rates. Although small temperature increases can raise the risk of malaria transmission, very high temperatures can be lethal to the mosquito and the parasite (Bradley 1993). Thus, where a localized warming falls within that mosquito species' survivability range is of critical importance.

The ability of a local environment to sustain a viable mosquito population is also a factor. Mosquitoes breed in shallow pools of standing water; too much water, and the environment becomes inhospitable to them. Thus, drought in tropical and subtropical wet and dry climates has been linked to enhanced opportunities for mosquito breeding. But very accurate precipitation forecasts will be needed to properly assess this component of the risk, and precipitation is a difficult variable to forecast in climate simulations.

As noted in the IPCC report (McCarthy et al. 2001), conditions for malaria transmission are so favorable at present that future climate change is unlikely to make the problem any worse. One concern is that, with regional warming, malaria will begin to invade higher-latitude regions that are considered to be immune (Epstein et al. 1998; McMichael et al. 1998). We must take care, however, to put these new cases in the proper historical context. Was malaria present during earlier, colder periods in our climate history? Has malaria existed in northern or higher-altitude climates? Asking and answering such questions reveals that observations of high-altitude malaria incidents in Madagascar and Costa Rica were actually not historically unusual, nor do they seem to be related to observed temperatures (Mouchet et al. 1998; Reiter 1998).

The complexity of the climatic component of malaria transmission alone can be illustrated by the following examples. Sufficient rainfall over relatively flat ground can produce mosquito-breeding pools sufficient to establish a population. But continued heavy rains can also remove mosquitoes from these pools. Drought can produce breeding pools in ephemeral streams but also cause the pools to evaporate. In some countries, drought may encourage people to capture rainwater in cisterns, thereby providing ideal breeding sites (Reiter 2001). Clearly, simple models cannot capture the variables necessary to develop accurate forecasts based on temperature and precipitation changes alone.

Although malaria and most other vector-borne diseases are commonly considered tropical diseases based on where they are currently endemic, climate has little to do with this pattern. Malaria was once very common throughout Europe, including all of Scandinavia, and the United States (figure 8.6). In the twentieth century, Archangel, Russia, just south of the Arctic Circle, experienced significant mortality from malaria. These occurrences were unrelated to yearly or interannual temperature changes. Malaria has essentially been eradicated from most middle and high-latitude countries because of major mosquito and plasmodium control efforts and housing changes by which people limit their contact with mosquitoes (Reiter 2000).

Despite the various caveats, given projected climate changes alone, it is likely the malaria occurrence would increase globally via spread into new

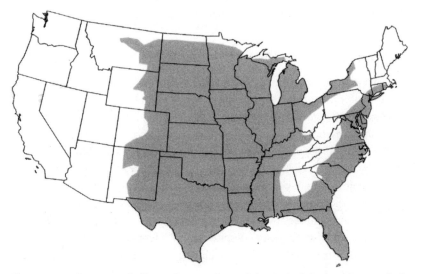

Figure 8.6.   Gray areas indicate those regions of the United States where malaria was endemic in the 1880s, a period when the climate was not unusually warm (Source: Reiter 1996).

regions (Rogers and Packer 1993); but the assumptions inherent in that statement preclude its usefulness. In 1977, 83 percent of the world's population was living in regions that were malarial at one time but in which malaria had already been, or was in the process of being, eradicated (Reiter 2000). But the eradication efforts have lost momentum and since then, malaria has begun to reappear in some of these areas. According to Reiter (2000):

> This rapid recrudescence has been attributed to population increase, forest clearance, irrigation and other agricultural activities, ecologic change, movement of people, urbanization, deterioration of public health services, resistance to insecticides and antimalarial drugs, deterioration of vector control operations, and disruptions from war, civil strife, and natural disasters. Claims that malaria resurgence is due to climate change ignore these realities and disregard history. For example, the many statements that recent climate change has caused malaria to ascend to new altitudes are contradicted by records of its distribution in 1880 to 1945.

The solutions to reduce malaria infection need not be costly. Simple use of screens and mosquito netting has a major impact on infection rates. Removal of prime mosquito breeding sites, such as unused rubber tires, also reduces the populations. Nevertheless, despite the success of these simple and inexpensive measures, it is likely that malaria (and other vector-borne diseases) will continue to predominantly impact underdeveloped nations.

Other Vector-Borne Diseases

Most of these comments regarding malaria are applicable to the transmission of other vector-borne diseases as well. Dengue and yellow fever have garnered significant attention, but research also continues on diseases transmitted by ticks, flies, and other vectors (McCarthy et al. 2001).

The importance of nonclimatic factors in disease transmission is illustrated by a now well-known example of dengue fever from 1995. In that year, a major dengue pandemic occurred in the Caribbean and Mexico, to the extent that more than two thousand cases of dengue surfaced in the small Mexican border town of Reynosa. Yet the entire state of Texas experienced only seven cases, despite nearly identical climatic conditions. Clearly, infrastructural rather than climatic factors are critical in dengue transmission rates.

Dengue and malaria have been linked to a large-scale multidecadal climate phenomenon known as El Niño. A period warming and cooling of ocean temperatures in the eastern tropical Pacific Ocean has been linked, via the ocean's interaction with the overlying atmosphere, to climate variability in many parts of the globe. El Niño (high eastern Pacific temperatures) and its cold counterpart, La Niña, have a fairly inconsistent periodicity of five to seven years. Because these oceanic changes are related to an atmospheric pressure oscillation in the southern hemispheric tropics (the Southern Oscillation), the combined ocean-atmosphere linkage is often referred to as the El Niño–Southern Oscillation (ENSO). Hales et al. (1996) observed that the number of dengue epidemics in the South Pacific were related to the timing of La Niña events. Similarly, Bouma and van der Kaay (1996) determined that malaria epidemics in Ceylon and India's Punjab province occurred in the wet monsoon season that followed dry El Niño years. They based this analysis on data from 1868 to 1943, a period prior to the introduction of insecticides.

While it is possible that El Niño has some minor influence on the likelihood of vector-borne disease transmission, the signal is so small relative to other factors, including even local temperature, rainfall, and humidity, that it is unlikely to provide much useful predictive power. For example, Bouma et al. (1994) invoke the following explanations of regional El Niño influences on malaria:

- In Sri Lanka: El Niño–related drought results in more standing water in rivers and larger mosquito populations.
- In Pakistan, Ethiopia, and Madagascar: High late-season temperatures induced by El Niño increase transmission rates.
- In the southeastern United States: "The historic unexplained epidemics in the south-eastern USA were reported between ENSO years, and may have been determined by lower temperatures during the ENSO events which temporarily restricted transmission in a period during which malaria was in decline due to other causes."

Although ENSO impacts vary regionally, the lack of a consistent understanding as to exactly what climatic aspects of ENSO actually influence disease transmission effectively eliminates any utility in using ENSO status for predictive purposes. ENSO alone does not cause regional climate changes— in some locations, it can modify wind flow, storm tracks, and so on in such a way that it can ultimately influence temperature, precipitation, and humidity. In most cases, these relationships are not consistent and vary quite markedly between ENSO events. Furthermore, little is known as to whether El Niños, La Niñas, or both, or neither are likely to increase in frequency in a greenhouse climate. Thus, efforts attempting to tie ENSO to disease transmission may exhibit some correlations but will fail when used in predictive models, owing to the lack of understanding of the physical linkages throughout the ocean-climate-weather-local disease transmission system cascade.

Climate is but one of many factors that influence the transmission of vector-borne diseases. Everything else being equal, the predicted climate changes would likely increase the spread of some of these diseases. But the forecasted temperature changes are relatively small, so major changes in habitats are unlikely. Ultimately, given the many other confounding socioeconomic factors related to disease transmission, it is likely that any real climate signals would be lost in the background noise. Independent of the climate change issue, good public health practice requires significant, ongoing, and consistent efforts to eradicate disease vectors and pathogens from the populations that are currently most susceptible to these diseases, and to maintain that eradication through consistent monitoring and reaction to renewed threats.

CONCLUSION

Climate has an impact on human life and livelihood. Aside from its obvious spheres of influence, such as agriculture, forestry, tourism, commodities, travel, storm impacts, and so forth, climate also can directly influence rates of human death and disease. In the summer, death rates are higher on hot or hot and humid days. Sequences of hot and humid days raise mortality rates. Declines in air quality in summer because of high ozone levels and other pollutants have been linked in some studies to higher death rates. Infectious diseases transferred to humans via insect vectors such as mosquitoes show a marked temperature dependence. In winter, death rates are universally higher, but the linkage to climate is, at present, weak. Communicable disease transmission is higher, possibly because of closer contact between individuals spending more time indoors during less pleasant weather. Winter air quality can be very poor under the influence of certain highly stable air masses that are more common in that season, but in most locations those weather conditions only persist for a few days.

Given the projections that temperature will rise in the future from increasing greenhouse gas concentrations, it is tempting to extrapolate current human impacts of climate forward in time to estimate future impacts. But this exercise is prone to error in many different ways. A typical presentation might state that we should expect an extra five thousand malaria deaths in Rwanda from each degree Celsius of warming above today's baseline value.

But those types of forecasts are usually based on an "everything else being equal" assumption that is frequently incorrect. An exceedingly large number of factors could potentially invalidate such a forecast. For example, the model's prediction for that region could be wrong. Ecological changes could result in shifts in key mosquito species. A new, highly effective antimalarial vaccine could be developed. Residents could make a concerted effort to eradicate the vector in infested areas using insecticides, to eliminate stagnant water pools, or to install screening and netting. Natural or human-driven habitat shifts could result in major reductions in the mosquito populations. In fact, all of those events have occurred historically in some regions. We could similarly develop a comparable list of changes that could make malaria outbreaks worse than predicted.

The fundamental problem is one of signal-to-noise ratio. In all the human impacts studied, the climate signal is very small relative to the noise, which in this case includes human behavior in addition to more predictable changes in the physical environment. The ability to make any forecast implicitly depends on the signal being sufficiently strong to be evident amidst the background variability. But human behavior, which specifically includes the conscious choice of individuals and societies to adapt to external changes, generates a very high level of background noise. The net result is a system that, at least for the time being, precludes the possibility of useful forecasts.

Nevertheless, planners and policymakers are faced with attempting to forecast a future based on models that have little accuracy and are based on assumptions that are inherently of low confidence. On the one hand, it is better to be prepared in spite of the uncertainty. Conversely, planning an incorrect course of action is actually worse than no plan at all. It is a simple truism to state that flexibility must be built into any decision-making process. But given a choice, for example, to commit limited monetary resources to develop a new dengue model or to buy window screens, the decision should be obvious.

In summary, developed nations are now, and in the future will likely remain, largely immune from major impacts of climate change on human health and well-being. These societies have the infrastructure and demonstrated ability to adapt to climate variability. Underdeveloped nations will likely be less successful. Yet the extent of human biophysical adaptations, though not well understood, will certainly be a factor in mitigating against major impacts. Furthermore, the climate changes likely to occur are small relative to the sensitivity of the systems they will purportedly influence. Given the inherent nature of living things to adapt, it is likely that the net impact of projected climate changes on human health will be very limited.

REFERENCES

Anderson, H. R., A. Ponce de Leon, J. M. Bland, J. S. Bower, J. Emberlin, and D. P. Strachan. 1998. Air pollution, pollens, and daily admissions for asthma in London, 1987–92. *Thorax* 53:842–48.

Applegate, W. B., J. W. Runyan Jr., L. Brasfield, M. L. M. Williams, C. Konigsberg, and C. Fouche. 1981. Analysis of the 1980 heat wave in Memphis. *J Am Geriatr Soc* 29:337–42.

Balling, R. C., Jr., P. J. Michaels, P. C. Knappenberger. 1998. Analysis of winter and summer warming rates in gridded temperature time series. *Clim Res* 9:175–81.

Bouma, M. J., H. E. Sondorp, and H. J. van der Kaay. 1994. Climate change and periodic epidemic malaria. *Lancet* 343:1440.

Bouma, M. J., and H. J. van der Kaay. 1996. The El Nino–Southern Oscillation and the historic malaria epidemics on the Indian subcontinent and Sri Lanka: An early warning system for future epidemics? *Tropical Med Intl Health* 1:86–96.

Bradley, D. J. 1993. Human tropical diseases in a changing environment. In *Environmental change and human health*. Edited by J. Lake, K. Ackrill, and G. Bock. London: CIBA Foundation.

Bridger, C. A., F. P. Ellis, and H. L. Taylor. 1976. Mortality in St. Louis, Missouri, during heat waves in 1936, 1953, 1954, 1955, and 1966. *Environ Res* 12:38–48.

Carey, M. J., and I. Cordon. 1986. Asthma and climatic conditions: experience from Bermuda, an isolated island community. *BMJ* 293:843–44.

Chan, N. Y., M. T. Stacey, A. E. Smith, K. L. Ebi, and T. F. Wilson. 2001. An empirical mechanistic framework for heat-related illness. *Clim Res* 16:133–43.

Danet, S., F. Richard, M. Montaye, S. Beauchant, B. Lemaire, C. Graux, et al. 1999. Unhealthy effects of atmospheric temperature and pressure on the occurrence of myocardial infarction and coronary deaths: A 10-year survey. The Lille-World Health Organization MONICA project (Monitoring Trends and Determinants in Cardiovascular Disease). *Circulation* 100:E1–7.

Davis, R. E., P. C. Knappenberger, W. M. Novicoff, and P. J. Michaels. 2002. Decadal changes in heat-related human mortality in the Eastern United States. *Clim Res* 22:175–84.

Davis, R. E., P. C. Knappenberger, W. M. Novicoff, and P. J. Michaels. 2003a. Decadal Changes in Heat-Related Human Mortality in the Eastern United States. *Intl J Biomet* 47:166–75.

Davis, R. E., P. C. Knappenberger, P. J. Michaels, and W. M. Novicoff. 2003b. Changing heat-related mortality in the United States. *Envir Health Perspect,* doi:10.1289/ehp.6336.

Davis, R. E., P. C. Knappenberger, P. J. Michaels, and W. M. Novicoff. 2004. Seasonality of climate-human mortality relationships in U.S. cities and impacts of climate change. *Clim Res* 26:61–76.

Dawson, K. P., J. Allan, and D. M. Fergusson. 1983. Asthma, air pollution, and climate: A Christchurch study. *NZ Med J* 96:165–67.

Dockery, D. W., F. E. Speizer, D. O. Stram, J. H. Ware, J. D. Spengler, and B. G. Ferris Jr. 1989. Effects of inhalable particles on respiratory health of children. *Am Rev Respir Dis* 139:587–94.

Dockery, D. W., C. A. Pope, III, X. Xu, J. D. Spengler, J. H. Ware, M. E. Fay, B. G. Ferris Jr., and F. E. Speizer. 1993. An association between air pollution and mortality in six U.S. cities. *N Eng J Med* 329:1753–59.

Donaldson, G. C., and W. R. Keatinge. 1997. Early increases in ischaemic heart disease mortality dissociated from and later changes associated with respiratory mortality after cold weather in southeast England. *J Epidemiol Comm Health* 51:643–48.

Donaldson, G. C., S. P. Ermakov, Y. M. Komarov, C. P. McDonald, W. R. Keatinge. 1998. Cold-related mortalities and protection against cold in Yakutsk, eastern Siberia: observation and interview study. *BMJ* 317:978–82.

Donaldson, G. C., R. S. Kovats, W. R. Keatinge, and A. J. McMichael. 2001. Heat- and cold-related mortality and morbidity and climate change. In *Health effects of climate change in the UK.* London: U.K. Department of Health.

Eng, H., and J. B. Mercer. 1998. Seasonal variations in mortality caused by cardiovascular diseases in Norway and Ireland. *J Cardiovascular Risk* 5:89–95.

Epstein, P. R., H. F. Diaz, S. Elias, G. Grabherr, N. E. Graham, W. J. M. Martens, et al. 1998. Biological and physical signs of climate change: focus on mosquito-borne diseases. *Bull Amer Met Soc* 79:409–17.

Epton, M. J., I. R. Martin, P. Graham, P. E. Healy, H. Smith, R. Balasubramaniam, I. C. Harvey, et al. 1997. Climate and aerallergen levels in asthma: A 12-month prospective study. *Thorax* 52:528–34.

Eurowinter Group. 1997. Cold exposure and winter mortality from ischaemic heart disease, cerebrovascular disease, respiratory disease, and all causes in warm and cold regions of Europe. *Lancet* 349:1341–46.

Frost, D. B., and A. Auliciems. 1993. Myocardial infarct death, the population at risk, and temperature habituation. *Int J Biometeorol* 37:46–51.

Gaffen D. J., and R. J. Ross. 1998. Increased summertime heat stress in the U.S. *Nature* 396:529–30.

Goldstein, I. F. 1980. Weather patterns and asthma epidemics in New York City and New Orleans, U.S.A. *Int J Biometeorol* 24:329–39.

Gover, M. 1938 Mortality during periods of excessive temperature. *Public Health Rep* 53:1122–43.

Greenberg, J. H., J. Bromberg, C. M. Reed, T. L. Gustafson, and R. A. Beauchamp. 1983. The epidemiology of heat-related deaths, Texas—1950, 1970–79, and 1980. *Am J Public Health* 30:130–36.

Guest, C. S., K. Willson, A. Woodward, K. Hennessy, L. S. Kalkstein, C. Skinner, and A. J. McMichael. 1999. Climate and mortality in Australia: Retrospective study, 1979–1990, and predicted impacts in five major cities in 2030. *Clim Res* 13:1–15.

Hajat, S., S. A. Goubet, and A. Haines. 1997. Thunderstorm-associated asthma: The effect on GP consultations. *British Journal of General Practitioners* 47:639–41.

Hajat, S., A Haines, S. A. Goubet, R. W. Atkinson, and H. R. Anderson. 1999. Association of air pollution with daily GP consultations for asthma and other lower respiratory conditions in London. *Thorax* 54:597–605.

Hales, S., P. Weinstein, and A. Woodward. 1996. Dengue fever epidemics in the South Pacific: Driven by El Nino Southern Oscillation? *Lancet* 348:1664–65.

Hegerl, G. C., and J. M. Wallace. 2002a. Influence of patterns of climate variability on the difference between satellite and surface temperature trends. *J Climate* 18:2412–28.

Hegerl, G. C., and J. M. Wallace. 2002b. Lapse rate trends and climate variability. *Bull Amer Meteorol Soc* 83:1293–94.

Henschel, A., L. L. Burton, L. Margolies, and J. E. Smith. 1969. An analysis of the heat deaths in St. Louis during July 1966. *Am J Public Health* 59:2232–42.

Ibald-Mulli, A., J. Stieber, H.-E. Wichmann, W. Koenig, and A. Peters. 2001. Effects of air pollution on blood pressure: A population-based approach. *Am J Public Health* 91:571–77.

Jones, T. S., A. P. Liang, E. M. Kilbourne, M. R. Griffin, P. A. Patriarca, S. G. Fite-Wassilak, et al. 1982. Morbidity and mortality associated with the July 1980 heat wave in St. Louis and Kansas City, Missouri. *J Am Med Assoc* 247:3327–31.

Kalkstein, L. S. 1991. A new approach to evaluate the impact of climate upon human mortality. *Environ. Health Perspec.* 96:145–50.

Kalkstein, L. S. 1993. Health and Climate Change: Direct impacts in cities. *Lancet* 342:1397–99.

Kalkstein, L. S., and J. S. Greene. 1997. An evaluation of climate/mortality relationships in large U.S. cities and the possible impacts of a climate change. *Environ Health Perspec* 105:84–93.

Kalkstein, L. S., P. F. Jamason, J. S. Greene, J. Libby, and L. Robinson. 1996. The Philadelphia hot weather-health watch/warning system: development and application, summer 1995. *Bull Amer Meteorol Soc* 77:1519–28.

Kalkstein, L. S., and R. E. Davis. 1989. Weather and human mortality: An evaluation of demographic and interregional responses in the United States. *Ann Assoc Amer Georgr* 79:44–64.

Katsouyanni, K., A. Pantazopoulou, G. Touloumi, I. Tselepidaki, K. Moustris, D. Asimakopoulos, et al. 1993. Evidence for interaction between air pollution and high temperature in the causation of excess mortality. *Arch Environ Health* 48:235–42.

Keatinge, W. R., S. R. K. Coleshaw, and J. Holmes. 1989. Changes in seasonal mortality with improvement in home heating in England and Wales from 1963–84. *Int J Biometeorol* 33:71–76.

Kilbourne, E. M. 1997. Heat waves and hot environments. In *The public health consequences of disaster.* Edited by E. K. Noji. New York: Oxford University Press.

Kilbourne, E. M., K. Choi, T. S. Jones, S. B. Thacker, and Field Investigation Team. 1982. Risk factors for heatstroke: a case-control study. *JAMA* 247:3332–36.

Kloner, R. A., W. K. Poole, and R. L. Perritt. 1999. When throughout the year is coronary death most likely to occur? A 12-year population-based analysis of more than 220,000 cases. *Circulation* 100:1630–34.

Kunst, A. E., C. W. N. Looman, and J. P. Mackenbach. 1993. Outdoor air temperature and mortality in the Netherlands: a time series analysis. *Am J Epidemiol* 137:331–41.

Langford, I. H., and G. Bentham. 1995. The potential effects of climate change on winter mortality in England and Wales. *Int J Biometeorol* 38:141–47.

Lanska, D. J., and R. G. Hoffmann. 1999. Seasonal variation in stroke mortality rates. *Neurology* 52:984–90.

Laschewski, G., and G. Jendritzky. 2002. Effects of the thermal environment on human health: An investigation of 30 years of daily mortality data from SW Germany. *Clim Res* 21:91–103.

Lerchl, A. 1998. Changes in the seasonality of mortality in Germany from 1946 to 1995: The role of temperature. *Int J Biometeorol* 42:84–88.

Lye, M., and A. Kamal. 1977. The effects of a heat wave on mortality rates in elderly patients. *Lancet* 1:529–31.

Lyster, W. R. 1976. Death in summer. Letter to the editor. *Lancet* 2:469.

Marmor, M. 1975. Heat wave mortality in New York City, 1949 to 1970. *Arch Environ Health* 30:130–36.

Martens, W. J. M., T. H. Jetten, and D. A. Focks. 1997. Sensitivity of malaria, schistosomiasis, and dengue to global warming. *Climate Change* 35:145–56.

McCarthy, J. J., O. F. Canziani, N. A. Leary, D. J. Dokken, and K. S. White (Eds.). 2001. *Climate change 2001: Impacts, adaptation, and vulnerability.* Cambridge: Cambridge University Press.

McGeehin, M. A., and M. Mirabelli. 2001. The potential impacts of climate variability and change on temperature-related morbidity and mortality in the United States. *Environ Health Perspec* 109:185–98.

McGregor, G. R. 1999. Winter ischaemic heart disease deaths in Birmingham, UK: A synoptic climatological analysis. *Clim Res* 13:17–31.

McGregor, G. R. 2001. The meteorological sensitivity of ischemic heart disease mortality events in Birmingham, UK. *Int J. Biometeorol* 45:133–42.

McMichael, A. J., J. Patz, and R. S. Kovats. 1998. Impacts of global environmental change on future health and health care in tropical countries. *Br Med Bull* 54: 475–88.

Michaels, P. J., P. C. Knappenberger, R. C. Balling Jr., and R. E. Davis. 2000. Observed warming in cold anticyclones. *Clim Res* 14:1–6.

Moolgavkar, S. H., E. G. Luebeck, T. A. Hall, and E. L. Anderson. 1995. Air pollution and daily mortality in Philadelphia. *Epidemiology* 6:476–84.

Mouchet, J., S. Manguin, J. Sircoulon, S. Laventure, O. Faye, and A. W. Onapa. 1998. Evolution of malaria in Africa for the past 40 years: Impact of climatic and human factors. *J Am Mosq Control Assoc* 14:121–30.

Nakai, S., T. Ioh, and T. Morimoto. 1999. Deaths from heat stroke in Japan: 1968–1994. *Int J Biometeorol* 43:124–27.

National Assessment Synthesis Team. 2000. Climate change impacts on the United States: the potential consequences of climate variability and change. U.S. Global Change Research Program, Washington, D.C.

Oeschsli, F. W., and R. W. Buechley. 1970. Excess mortality associated with three Los Angeles September hot spells. *Environ Res* 3:277–84.

Oke, T. R. 1987 *Boundary layer climates.* Cambridge: Cambridge University Press.

Palecki, M. A., S. A. Changnon, and K. E. Kunkel. 2001. The nature and impacts of the July 1999 heat wave in the Midwestern United States: Learning from the lessons of 1995. *Bull Am Meteor Soc* 82:1353–67.

Pell, J. P., and S. M. Cobbe. 1999. Seasonal variations in coronary heart disease. *QJM* 92:689–96.

Platts-Mills, T. A. E., and M. C. Carter. 1997. Asthma and indoor exposure to allergens. *N Eng J Med* 336:1382–84.

Pope, C. A., III, and D. W. Dockery. 1992. Acute health effects of PM10 pollution on symptomatic and asymptomatic children. *Am Rev Respir Dis* 145:1123–28.

Pope, C. A., III, D. W. Dockery, J. D. Spengler, and M. E. Raizenne. 1991. Respiratory health and PM_{10} pollution. *Am Rev Respir Dis* 144:668–74.

Pope, C. A., III, and L. S. Kalkstein. 1996. Synoptic weather modeling and estimates of the exposure-response relationship between daily mortality and particulate air pollution. *Environmental Health Perspectives* 104:414–20.

Reiter, P. 1996. Global warming and mosquito-borne disease in the USA. *Lancet* 348:662.

Reiter, P. 1998. Global warming and vector-borne disease in temperate regions and at high altitude. *Lancet* 351:839–40.

Reiter, P. 2000. From Shakespeare to Defoe: Malaria in England in the Little Ice Age. *Emerging Infectious Diseases* 6:1–11.

Reiter, P. 2001. Climate change and mosquito-borne disease. *Environ Health Perspectives* 109:141–61.

Rogers, D. J., and M. J. Packer. 1993. Vector-borne diseases, models, and global change. *Lancet* 342:1282–84.

Rogot, E., P. D. Sorlie, E. Backlund. 1992. Air-conditioning and mortality in hot weather. *Am J Epidemiol* 136:106–16.

Rosenstreich, D. L., P. Eggleston, M. Kattan, D. Baker, R. G. Slavin, P. Gergen, H. Mitchell, et al. 1997. The role of cockroach allergy and exposure to cockroach allergen in causing morbidity among inner-city children with asthma. *N Engl J Med* 336:1356–63.

Samet, J. M., F. Dominici, F. C. Curriero, I. Coursac, and S. L. Zeger. 2000. Fine particulate air pollution and mortality in 20 U.S. cities, 1987–1994. *N Engl J Med* 343:1742–49.

Samet, J. M., S. Zeger, J. Kelsall, J. Xu, and L. Kalkstein. 1998. Does weather confound or modify the association of particulate air pollution with mortality? *Environ Res* 77:9–19.

Schuman, S. H. 1972. Patterns of urban heat-wave deaths and implications for prevention: Data from New York and St. Louis during July 1966. *Environ Res* 5:59–75.

Schuman, S. H., C. P. Anderson, and J. T. Oliver. 1964. Epidemiology of successive heat waves in Michigan in 1962 and 1963. *J Am Med Assoc* 180:131–36.

Semenza, J. C., C. H. Rubin, K. H. Falter, J. D. Selanikio, W. D. Flanders, H. L. Howe, and J. L. Wilhelm. 1996. Heat-related deaths during the July 1995 heat wave in Chicago. *N Eng J Med* 335:84–90.

Sheridan, S. C., and T. J. Dolney. 2003. Heat, mortality, and level of urbanization: Measuring vulnerability across Ohio, USA. *Clim Res* 24:255–65.

Simonsen, L., M. J. Clarke, G. D. Williamson, D. F. Stroup, N. H. Arden, and L. B. Schonberger. 2001. The impact of influenza epidemics on mortality: Introducing a severity index. *Am J Pub Health* 87:1944–50.

Smoyer, K. E., L. S. Kalkstein, J. S. Greene, and H. Ye. 2000. The impacts of weather and pollution on human mortality in Birmingham, Alabama, and Philadelphia, Pennsylvania. *Int J Climatol* 20: 881–97.

Steib, D. M., R. T. Burnett, R. C. Beveridge, and J. R. Brook. 1996. Association between ozone and asthma emergency department visits in Saint John, New Brunswick, Canada. *Environ Health Perspectives* 104:1354–60.

Ware, J. H. 2000. Particulate air pollution and mortality: Clearing the air. *N Eng J Med* 343:1798–99.

9

POSSIBLE EFFECTS
OF SOLAR VARIABILITY
ON THE EARTH'S ECOSYSTEMS

Sallie Baliunas

The sun's electromagnetic radiation, magnetic fields, and particle flux vary on many spatial and temporal scales, making the sun one of several possible exo-terrestrial influences on the earth's environment. Over periods of decades to centuries—a relevant scale for comparison to the air's increased concentration of greenhouse gases—the sun's changes are especially marked in thermal, chemical, and dynamical changes in the middle and upper terrestrial atmosphere.

As for temperature trends in the lower atmosphere, a straightforward, presumed mechanism of solar–terrestrial influence would be an interdecadal increase in total solar irradiance, or the sun's energy output integrated over all frequencies of its electromagnetic spectrum. The measured change in total solar irradiance, made on a near-daily basis by a series of satellites since 1978, is relatively small—approximately 0.1–0.15 percent over a decade. Computer simulations of climate, encompassing multiple parameters and their complex, nonlinear dynamical interactions, including the increased concentration of atmospheric greenhouse gases, imply only a small contribution from the observed total solar irradiance change to the recent surface warming trend.

Thus, the U.N. Intergovernmental Panel on Climate Change's latest assessment on climate science (Houghton et al. 2001) stated that total solar irradiance change insufficient to explain surface temperature warming of the second half of the twentieth century.

However, recent research reveals surprisingly good correlation between measures of solar variability and local terrestrial ecosystem change suggesting sun–climate influences on periods of decades to centuries and longer. If such observed environmental variations were to be explained solely by solar total irradiance changes, they would be much larger over periods of decades to centuries than has been suspected. Alternatively or additionally, terrestrial processes such as ocean–atmosphere interactions may amplify small changes in total solar irradiance or specific aspects of solar change (e.g., variations of the ultraviolet portion of the sun's electromagnetic spectrum or the sun's highly variable fast-moving particle output). Finally or additionally, the flux to the earth's upper atmosphere of energetic particles produced primarily by galactic supernovae, called cosmic rays, changes as the sun's varying magnetism modulates the cosmic ray flux. That has led to speculation that cosmic rays contribute to environmental change, perhaps through the intermediary of terrestrial clouds. Proposed solar and cosmic ray processes that seek to explain the observed, apparent exo-terrestrial influences seen in environmental reservoirs cannot yet be accurately quantified. Hence, it remains difficult to apportion ecosystem change among exo-terrestrial and other natural effects, plus anthropogenic influences such as the enhanced greenhouse effect, black carbon emission, and landscape modification.

Incident sunlight is the primary energy input to the climate system. But the energy emitted by the sun is not constant, a fact that has provoked much speculation as to whether some of the temporal evolution of climate response results from solar variability. This question has acquired more importance as human activities alter both the surface of the earth and its atmosphere.

Regarding the solar influence on globally averaged surface temperature, the *Summary for Policymakers* (SPM) carried in Working Group I's science assessment (Houghton et al. 2001) of the U.N. Intergovernmental Panel on Climate Change notes, "Simulations of the response to natural forcings alone (i.e., the response to variability in solar irradiance and volcanic eruptions) do not explain the warming in the second half of the 20th century" (*Summary for Policymakers*, 10).

However, Working Group I states important caveats to that conclusion: "Because of the large uncertainty over the absolute value of [Total Solar Irradiance, or TSI] and the reconstruction methods, our assessment of the 'level of scientific understanding' is 'very low'" (382). Further, "knowledge of solar radiative forcing is uncertain, even over the 20th century and certainly over longer periods" (382). Working Group I's report (382–85) does consider possibilities of the sun–climate influence, beyond that of total solar irradiance change, including enhanced terrestrial response to variable ultraviolet solar irradiance or cosmic ray flux, and correctly notes, "We conclude that the mechanisms for the amplification of solar forcing are not well established" (385). Thus, the SPM statement, emphasizing the impact of total solar irradiance alone, is incomplete, in part because, as Working Group I concludes, the sun's environmental influence remains uncertain. Since Working Group I's report was compiled, Stott et al. (2003), for example, have argued that solar forcing may have been underestimated in most climate simulations. Specifically, simple linear extrapolations of the total solar irradiance may need to be stretched—by a factor of one (no stretching necessary) to six, depending on other factors in the climate simulation, to explain the early twentieth century's surface temperature trend, when the effect of increased atmospheric concentration of greenhouse gases was relatively low. The SPM statement also leaves unexplained the body of impressive correlations observed between the solar magnetism seen in records of cosmogenic isotopes and ecosystem parameters studied in terrestrial environmental reservoirs. That lack of knowledge of the causes of the correlations—beyond just those of total irradiance observed over two decades—on time scales of decades to centuries means that their impacts on climate cannot be reliably reproduced by simulations.

Perhaps implicitly reflecting the caveats of Working Group I, SPM (figure 3, p. 8) correctly assigns its poorest confidence level—"very low" confidence—to knowledge of solar radiative forcing. Quantitatively, the sun is assigned an equivalent radiative forcing that is small (less than 0.5 Wm^{-2} between 1750 and 2000), with an even smaller uncertainty, under the assumption that the total solar irradiance is the dominant mechanism for solar terrestrial influence. Further, TAR SPM concludes that surface "warming over the past 100 years" "was unusual and unlikely[7] to be entirely natural in origin" (10), where the superscript 7 denotes a footnote defining "unlikely" as "10–33% chance" that a result is true, based on "judgmental estimates of confidence" (2). That mix of quantitative and subjective numbers appears despite poor confidence, meager quantitative knowledge about solar and other exo-terrestrial factors, and the correlations that have been observed linking solar magnetic change to local terrestrial ecosystem change on decadal to millennial scales. Moreover, potentially important anthropogenic influences, including landscape alteration (Chase et al. 2001) and biological responses to the increase in the air's carbon dioxide concentration, have only crudely if at all been incorporated in the simulations (e.g., Pielke 2002).

ESTABLISHING A VARIABLE SUN

Solar–terrestrial influences can be studied in three broad and interrelated ways: first, by trying to understanding the physical principles and constructing ab initio models of the sun and terrestrial environment. Even in the absence of success in achieving ab initio depictions, semiempirical models have helped to define limits of change and its causes. A second approach examines environmental indicators where not only terrestrial change but also coincident solar change is evident. Records from which far-past solar change are estimated are by-products of the interactions of cosmic rays with the terrestrial upper atmosphere. While the flux of cosmic rays to the upper atmosphere is modulated by solar magnetic variability, some researchers have argued for terrestrial influences on the environment by way of cosmic rays themselves; thus, the ecosystem impacts may be more broadly termed exo-terrestrial until their physical mechanisms have been established. A third approach looks to quantify the time scales and amplitudes of relevant solar change by considering closely related phenomena to the sun's seen in other stars. Because of the enormous scope of the topic, this work gives limited background and highlights environmental change on periods of centuries to millennia; interested readers should consult more comprehensive reviews beginning with, for example, Hoyt and Schatten (1997) plus Soon and Yaskell (2004).

Naked-eye sunspot sightings—a visible indication that the sun changes—date back over two millennia in China and Greece, with systematic observations recorded in China through much of that period (Sarton 1947; Bray and Loughhead 1964; Soon and Yaskell 2004). Because ecclesiastical authority in western post-medieval Europe had declared the sun to be a perfect and unblemished celestial orb, the idea of recording the dark spots, or maculae, required considerable rebellion, as in the case of Galileo. So culturally pervasive was the idea of solar perfection that on May 18, 1607, the skilled observer Johann Kepler misconstrued a sunspot for a transit of Mercury. But by the early seventeenth century the development of the telescope for astronomical use helped shatter the notion of the sun as a perfect crystalline celestial sphere, and so began sustained telescopic observations by Galileo in Italy, Johannes Goldsmid (under his Latinized name, Fabricius) and his father, David, in Holland, Thomas Harriot in England, and Christopher Scheiner in Germany (Bray and Loughhead 1964; Hoyt and Schatten 1997; Soon and Yaskell 2004).

The great astronomer Sir William Herschel (1738–1822) noted that the number of sunspots apparent on the disk of the sun rose and fell over the years, and studied their possible connection to changes in climate. (It was not until more than forty years later that Henrich Schwabe [1843], an amateur astronomer, discovered the decadal cycle of the increase and decrease of the number of sunspots.) Herschel (1801) hypothesized that periods of numerous surface spots "may lead us to expect copious emission of heat and therefore mild seasons," contrasted with "severe seasons" and "spare emission of heat" at times of few sunspots. Herschel examined five long periods when sunspots were few and turned to wheat prices in England as a local climate indicator under the expectation that wheat harvest would be adversely affected in periods of low sunspot activity. Herschel did find scanty evidence to uphold his supposition, and fully explained the limits of the available information. Nonetheless, Herschel's careful work was severely criticized, for example, by Lord Brougham (1778–1868) who pompously pronounced it a "grand absurdity" as incredible as "Gulliver's voyage to Laputa" (Brougham 1803).

George Ellery Hale (1908) discovered with the newly built sixty-foot Solar Tower at Mount Wilson Observatory that sunspots consist of arched magnetic fields with a strength of approximately 0.2 to 0.3 T. A modern record of the number of sunspots, or solar surface magnetism, going back to the early seventeenth century, is shown in figure 9.1. Visual inspection reveals that the eleven-year sunspot cycle is an approximate description because the cycle averages one decade in length, with the length of a cycle ranging from eight to fifteen years. More properly it should be called

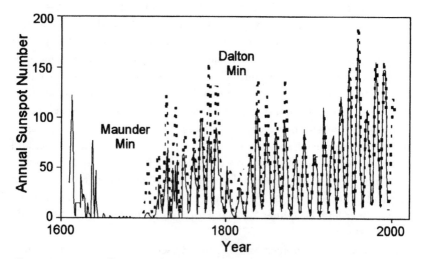

Figure 9.1.   Annually averaged sunspot number, a measure of solar magnetic activity (dotted line plotted from data available at ftp://ftp.ngdc.noaa.gov/stp/solar_data/sunspot_numbers/yearly.plt; solid line plotted from data available at ftp://ftp.ngdc.noaa.gov/stp/solar_data/sunspot_numbers/group_sunspot_numbers/yearrg.dat).

decadal variability.[1] Peak cycle intensities have generally swung upward since the beginning of the record in the early seventeenth century, although not monotonically. There are notable low periods, for example, the interval marked "Maunder Minimum," named in honor of E. W. Maunder (1890, 1894; Soon and Yaskell 2004), who reintroduced and helped popularize the period of sustained low sunspot counts (c. 1640–1720), with sunspots appearing predominantly in the sun's southern hemisphere. Other magnetic low periods are the brief, early nineteenth century's Dalton Minimum (Eddy 1976), and prior Wolf (sixteenth century), Spörer (thirteenth and fourteenth centuries), and Oort (eleventh century) minima.

Explaining solar magnetism remains an outstanding challenge in astrophysics. Parker (1955a, b; followed notably by Steenbeck et al. 1966; Leighton 1969; see Parker 1994) established the fundamentals of one class of models of solar surface magnetism, based on the α-Ω dynamo. The dynamo, drawing on the sun's internal turbulent convective motions generated in the outer 30 percent of the solar radius, converts a fraction of the kinetic energy to magnetic energy. Field lines are transported and twisted as they rise to the surface, where they erupt as magnetically active structures like sunspots, faculae, and active regions. Because the fluid is highly conductive, magnetic fields and energy couple nonlinearly back to the fluid, creating a complex, nonlinear, and regenerative process.

The sun's magnetic cycle begins when a globally extensive, weak poloidal field is stretched by nonuniform rotation, sweeping field lines into an azimuthal configuration that strengthens as the sun rotates. Helical motions arising from convection create small loops of buoyant magnetic field that erupt at the sun's midlatitudes at the beginning of the cycle, subsequently erupting at lower latitudes as the magnetic field strengthens toward the peak of the cycle. Persistent subsurface diffusion finally succeeds in relaxing the field and marks the cycle, beginning anew with a weak, poloidal field. Additionally, there is a reversal of polarity of the large-spatial scale expression of the magnetic field from one cycle to the next: the leading sunspot of pairs in the northern hemisphere tend to show one polarity during a cycle, while the other polarity obtains in the leading spots of the southern hemisphere. That pattern reverses at the next cycle, making a twenty-two-year magnetic polarity cycle (Hale and Nicholson 1925).

In summarizing the successes and deficiencies resulting from the application of dynamo models, Weiss and Tobias (2000) and Petrovay (2000) conclude that no model approach adequately describes the body of reasonably well-observed and relevant solar properties. Understanding the tachocline, the seat of the interaction of magnetism and convective motions, should be a fruitful step forward (Charbonneau et al. 1999), as should continued extraction of solar periodicities from observations, exemplified by the riches gotten from recent paleoecological information on millennial scales.

As already noted, solar variations are not regular, and numerical techniques have advanced to describe the periodicities. Departing from traditional Fourier calculations, which detect persistent sinusoidal phenomena and are ill suited to analysis of the sunspot time series (or any other known solar magnetism series), wavelet analysis has been used to describe the time evolution of solar periodicities. From a record of sunspot groups from 1610 to 1994, the wavelet analyses of Frick et al. (1997) reveal a primary eleven-year periodicity, plus the longer Gleissberg (1944) periodicity of approximately one hundred years. The eleven-year cycle disappears in Frick et al.'s analysis in the period 1650–1680, within the window of the Maunder Minimum (c. 1640–1720), when solar magnetism was relatively weak compared to the twentieth century. Also using wavelet analysis, but for the sunspot record from 1700 to 2002, Le and Wang (2003) find not only the 11-year and 100-year periodicities, but also a 56-year periodicity that is especially strong between 1725 and 1850 and weak outside that interval. Ram and Stolz (1999) report ~11-year, 100-year (Gleissberg), and ~200-year modulations of dust in layers in an ice core from Greenland—the first suggestion that familiar, modern solar decadal and interdecadal variations pertain as far back as 90,000 years.

Thus, the relative amplitudes of the dominant periodicities found for the sun's magnetism vary in time, emphasizing its complexity. No theory or model yields successful explanation or forecasts of solar magnetism and its variability over decadal, interdecadal, or longer periods.

EARLY-NOTED SOLAR MAGNETIC-TERRESTRIAL INTERACTIONS

Electrical currents in the terrestrial atmosphere, originating in solar transient magnetic phenomena, generate geomagnetic fluctuations, first noted by O. P. Hiorter in 1741 (Brekke and Egeland 1983). In the latter half of the nineteenth century geomagnetic activity had been found to vary with the decadal sunspot cycle (Sabine 1852; Moldwin 2004). Carrington (1860) and Hodgson (1860) visually observed flares that subsequently produced on September 1, 1859, what seems to have been the greatest auroral storm in recent centuries.

Magnetic fields are now known to dominate the transfer of energy through the solar atmosphere and then reach outward to permeate the solar system as they are carried by the magnetized solar wind (Parker 1958). Eruptive processes like flares or coronal mass ejections loft fields and particles that may interact with the geomagnetic, other planetary or interplanetary magnetic fields, plus planetary bodies and atmospheres. Extending outward, the sun's magnetic wind halts only at the heliopause at a distance presently estimated to be approximately 153 to 158 astronomical units (Gurnett et al. 2003), as detected by Voyager 1 spacecraft, launched September 1977.

Highly variable, solar-driven auroral and geomagnetic fluctuations per se seem energetically too weak (and too ephemeral) to lead to significant and direct low-atmosphere terrestrial temperature variability over decades or longer.

Cosmogenic Isotopes and Millennia-Scale Records of Solar Magnetism

By the early 1960s another important influence of the sun's magnetic field on the terrestrial environment—and an influence that leaves a voluminous record of past solar change—had been realized. Radiocarbon is formed when slow neutrons produced by cosmic ray by-products are captured by nitrogen atoms in the upper terrestrial atmosphere; quickly oxidizing, the radiocarbon may be deposited into biogeological reservoirs—tree growth rings, loess, sediments, coral, speleothems, and so on. Additionally, the isotope ^{10}Be is formed by cosmic ray by-products with atmospheric nitrogen or oxygen atoms, and sometimes precipitates into ice layers in glaciers or ice sheets.

The cosmic rays impinging on the solar system are nucleons, consisting mostly of protons but with all elements in the periodic table up to uranium represented, moving at a spectrum of relativistic speed with energies observed as high as $\sim 10^{20}$ eV. Except for the highest-energy cosmic rays, which are relatively few and extragalactic in origin, the nucleons have been accelerated to their great speeds by galactic supernovae and their remnants. Because of the average rate of galactic supernova—approximately two per century—and the lifetime of their remnants, the cosmic ray flux bombarding the solar system is relatively constant over periods of millions of years. On reaching the solar system, the cosmic rays are subject to interactions with the local interplanetary magnetic field, which is dominated by the sun's magnetic fields that are varying over periods of seconds to millennia and longer, and thus modulate the flux of particles as a function of their energies. For example, the flux of 100 MeV protons to the earth has been estimated to vary by as much as a factor of ~100 compared to a factor of ~3–5 for the flux of 2 GeV protons during the past three hundred years as a result of solar magnetic variability, according to analysis of 10Be concentration in the Dye-3 Greenland coring (McCracken et al. 2004). The geomagnetic field and its variations also influence the cosmic ray flux incident on the atmosphere, and environmental processes—climate, biological, geological—affect the amount of cosmic ray by-products transported and deposited to ecological reservoirs. All those factors complicate the extraction of a signal of past solar magnetism. But the work has the potential to yield solar magnetic activity records well in excess of the length of the four-hundred-year telescopic record of sunspot numbers because the isotopes are relatively long-lived, 5,730-year for radiocarbon and 1.5 million-year for ^{10}Be (see Beer 2000).

Owing to its residence time in the air and ocean reservoirs of several decades (~20–60 years or longer), radiocarbon production records have yielded information mainly for interdecadal and longer periods. Stuiver (1961, 1965) showed that for radiocarbon fluctuations in the eighteenth and nineteenth centuries, periods of low radiocarbon content in tree cellulose corresponded to periods of sustained high sunspot number and vice versa. The shorter atmospheric residence time of ^{10}Be of a few years allows its concentration in ice cores to be directly related to cosmic ray fluxes measured by neutron counting monitors and solar magnetism through the sunspot record. Uncertainties in estimating cosmogenic isotopic production, transport, and deposition on the way toward gauging solar magnetism can be ameliorated by comparing common periodicities inferred from the two different cosmogenic records. One such record of solar magnetism over the last millennium combined from the ^{14}C and ^{10}Be records from a Greenland ice core and shown on the radiocarbon production scale is shown in figure 9.2 (from Bard et al. 1997). The record of ^{10}Be from a different location, the South Pole, is also shown for comparison in figure 9.2 (Raisbeck et al. 1990; Raisbeck and Yiou 2004). Both records indicate high levels of solar magnetism nine hundred years ago, followed by periods of extremely low solar magnetism, and a return to high solar activity in the twentieth century.

Recently, Usoskin et al. (2004) transformed the ^{10}Be fluctuations to the scale of sunspot number in order to extend the historical record of sunspots back to approximately A.D. 850. They concluded that the latter half of the twentieth century has seen the period of highest sunspot number in the past 1,150 years. However, uncertainties in the steps in the construction

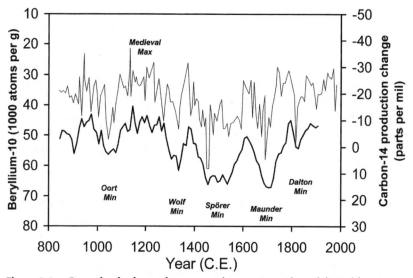

Figure 9.2. Records of solar surface magnetism or magnetic activity, with extrema of interdecadal solar magnetic variability labeled, constructed from radiocarbon and beryllium-10 measurements from a Greenland ice core (thick line), shown on fractional radiocarbon scale, which has been inverted so that solar activity increases upward (Bard et al. 1997); and from beryllium-10 measurements from a South Pole ice core (thin line) (Raisbeck et al. 1990; Raisbeck and Yiou 2004).

of an early sunspot record, as well as deficiencies in sunspot number as a physical quantity—for example, its limited dynamic range—leave open the question of the period of greatest number of sunspots (Raisbeck and Yiou 2004). As described above, the period of elevated magnetic activity about 900 years ago has been well documented directly from 14C and 10Be records; there is little doubt that the sun's magnetism was extremely strong between 900 years ago and the late nineteenth century (Raisbeck and Yiou 2004) then continuing through the twentieth century.

Taken together, sunspot (and, in an even less quantitative way, auroral sightings) and cosmogenic isotope records indicate fundamental solar variations of approximately a decade (Schwabe cycle), a century (Gleissberg cycle), 200 years (deVries or Suess cycle), and 2100–2400 years (Bray and Loughhead 1964; Suess 1968; Eddy 1976; Stuiver et al. 1991; Damon and Sonett 1991; Bard et al. 1997; Beer 2000; see also Soon and Yaskell 2004). The ~200-year periodicity may coincide with the appearance of the Dalton, Maunder, Spörer, and Wolf solar magnetic minima, and also appears between 25,000 and 50,000 years ago during the last glacial period (Wagner et al. 2001). Information about the century and millennial variations, or variations from tens of thousands of years ago, come from the cosmogenic isotope records, underscoring the importance of the records for studying solar magnetism and its variability.

As with decade- and century-scale variations, solar millennial variations are expected to occur without strict or great regularity.[2] However, peaks cited in spectral analyses to a relatively high precision should not always be interpreted as accurate periodicities produced by the sun, only the precise identification of power spectrum peaks. That distinction between the detection of peaks and assertion of source periodicities in the sun should be borne closely in mind when trying to identify solar impacts on terrestrial systems, which would filter solar variations through nonlinear, dynamic biogeological systems. Discussion of coincident cosmogenic and terrestrial changes continues below.

TOTAL SOLAR IRRADIANCE CHANGES

Although solar magnetism per se is thought to produce little significant climate response, solar magnetism has historically been closely associated with a primary suspect of terrestrial change, namely, total solar irradiance variability on interdecadal scales. Still called the solar constant, the total solar irradiance is the amount of light energy per second incident on a square meter of the earth's atmosphere at a distance from the sun of one astronomical unit, and integrated over all wavelengths.

Direct attempts to measure the solar constant in the past could only gather the portion of the irradiance accessible from ground-sited facilities and were made by Claude Pouillet and John Herschel in the early nineteenth century. Little could be found in the way of fluctuations in the solar constant in those early observations.

Samuel Pierpont Langley, the third secretary of the Smithsonian Institution, persuaded Congress to establish the Smithsonian Astrophysical Observatory in 1890. With his assistant Charles Greeley Abbot (who later became the institution's fifth secretary), Langley established a solar constant monitoring program that lasted approximately five decades. Abbot, who led the program for many years, recalled Langley's motivation for measuring the solar output: it would solve "the fundamental problem of meteorology, nearly all whose phenomena would become predictable, if we knew both the original quantity and kind of [solar] heat" (DeVorkin 1998).

Abbot's careful work led to an improved value of the solar constant. Influenced by meteorologists H. H. Clayton and A. Lawrence Rotch,

Abbot then turned to the question of a variable solar output over decades that might influence terrestrial climate and weather.

Abbot faced difficulties working near the bottom of the atmosphere. Not only were frequencies of the solar electromagnetic spectrum beyond the visible and near infrared absorbed, but also variable cloudiness and sky conditions could not be estimated well enough to correct accurately for the unmeasured components of solar energy output. Abbot's invaluable efforts to improve instruments, techniques, and observing stations by moving them to high mountaintops produced results too imprecise to measure solar irradiance change, which is subtler in amplitude than had been expected. In retrospect, Abbot found no convincing, quantitative link between climate and solar irradiance changes (although there were claims of such; see Hoyt and Schatten 1998).

Hale, too, had been interested in the solar–terrestrial influence, and wished Langley's program to be sited under excellent atmospheric conditions. Hale had founded Mount Wilson Observatory in part because he saw the cosmos with integrated processes, by analogy to the overarching concept of Darwinian evolutionary theory in biology. The theme of cosmic evolution struck Hale (1905), "For the Sun is a star, comparable in almost every respect with many other stars in the heavens, and rendering possible, through an intimate knowledge of its own phenomena, the solution of some of the most puzzling questions in the general problem of stellar evolution" (127–28).[3]

In the late twentieth century the solar constant's inconstancy was finally bared with satellite detectors that could measure the subtle variations of the total solar irradiance (Willson and Hudson 1988; Hickey et al. 1988), including approximately 0.15 percent amplitude over a decade and positive correlation with the sunspot cycle. Variations in total solar irradiance owe to several physical phenomena occurring on different time scales: p-mode oscillations clustered near five minutes; subsurface convective energy flow that appears on the surface as granulation over several tens of minutes; the growth and decay of sunspots over a few days; the growth and decay of bright magnetic surface features such as faculae over tens of days; and the sunspot or decadal magnetic variation.

The positive correlation between decadal magnetism and solar irradiance changes has been modeled as changing fluxes as a result of different lifetimes and variable coverage by solar surface magnetic features (e.g., Pap 1997). Very crudely, at sunspot maximum, the prevalence of spots, darker and cooler than the surrounding magnetically quiet photosphere, tend to reduce the instantaneous total irradiance. At nearly the same pacing, magnetically active features (e.g., faculae and the network) brighten the irradiance by increasing in number and surface coverage. In the net, it is the variation in the fluxes of bright magnetic features that tends to dominate the decadal irradiance curve, causing the positive correlation with magnetic activity. Quantitatively ascribing various solar components and their variability to explain total irradiance change remains for future work (Fontenela et al. 1999, 2004; Unruh et al. 1999; White et al. 2000; Foukal 2002; Ortiz et al. 2002; Fox et al. 2004), especially over interdecadal periods. On century and longer periods, there are no reliable models of solar components and their irradiance contributions.

As for solar variability from cycle to cycle, Willson and Mordvinov (2003) recently transformed total solar irradiance measurements made with various satellite instruments (ERB, ACRIM1, ACRIM2, ACRIM3, and VIRGO) in different periods to construct a record going back to 1978 (figure 9.3). As noted above, the full amplitude between maximum and minimum of the activity cycle is approximately 0.15 percent. Of interest is that the values of the total solar irradiance from sunspot minimum of 1986 to that of 1996 trended upward approximately 0.04 percent, suggesting a significant, recent interdecadal trend in total solar irradiance. However, one difficulty with the comparison of the different total solar irradiance data sets

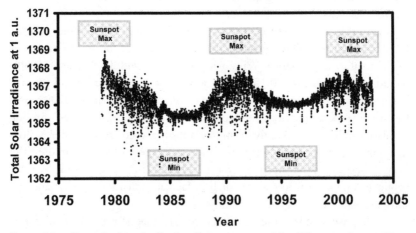

Figure 9.3. Record of total solar irradiance, measured by different solar satellites, and intercalibrated by Willson and Mordvinov (2003). Data are the NNAVA3 compilation (wherein ACRIM2 results from 1996 to 1998 are replaced with VIRGO results). The trend between sunspot minima is +0.044 percent (data are from ftp://ftp.ngdc.noaa.gov/stp/solar_data/solar_irradiance/acrim3/acrim3composite_nnava3.txt).

is that their absolute calibration differs by about 0.7 percent, a much larger factor than the relative amplitudes of physical phenomena (e.g., Fröhlich 2000). To quantify interdecadal total solar irradiance may require dedicated satellite monitoring.

While total solar irradiance change at first seems a good and direct candidate as a factor in climate change, estimates of the amplitude of total solar irradiance extrapolated to interdecadal and longer periods, and considered within present-day climate simulations, rules out total solar irradiance as a dominant climate response, especially in the latter half of the twentieth century, when the atmospheric concentration of greenhouse gases has increased substantially (e.g., Soon et al. 1996; Shindell et al. 1999; Rind 2002). However, uncertainties are still significant because of the limitations arising from the duration and precision of direct total irradiance measurements, models of the solar components contributing to irradiance change, and knowledge of their climatic responses, especially over interdecadal scales. As noted below, observed local ecosystem changes that correlate well with solar magnetism over centuries to millennia are reopening the question.

SOLAR LUMINOSITY CHANGES

The decadal changes in total solar irradiance over the last two sunspot cycles are observed in a restricted viewing angle as the irradiance is intercepted by satellite only within the earth's orbit around the sun. Thus, little information at other angles, for example, the irradiance emitted by the solar polar areas, is captured, but in terms of terrestrial impact, the view angle of satellite observatories is appropriate.

The observed total solar irradiance measurements have been shown to be actual luminosity changes from the sun, at or near the surface (Kuhn et al. 1988; Kuhn 2004), resulting from the storage and release of energy in the sun's subsurface convective zone and magnetic fields. A formal distinction is made here on the origin of luminosity changes based on their time scale. The variability on millennial and shorter scales considered here arises from surface and subsurface processes that temporarily store, then release, energy on those scales. Over periods of tens of millions of years and longer, the solar core's properties vary as fusion proceeds over the main sequence lifetime (the phase of predominantly proton–proton and ^{3}He–^{3}He fusion reactions in the core, a phase that lasts approximately ten Gyr).

Theoretical studies of the interior structure of the sun lead strongly to expectations of solar luminosity change over eons. Schwarzschild et al.

(1957) early noted that the solar luminosity should increase as a consequence of the steadily decreasing fraction by mass of hydrogen in the core as helium ash accumulates and increases the core density, core temperature and pressure, and hence, luminosity. Present-day solar interior models (see, e.g., Bahcall et al. 2001) predict a monotonic increase in the sun's luminosity of about 48 percent between the Zero Age Main Sequence and now (i.e., an age of approximately 4.6 Gyr). Those models have been refined by estimates of the interior sound speed derived from the inversion of superb helioseismology measurements, besides observable surface properties like mass, luminosity, and radius. One important, associated matter deserving comment is that even such well-grounded solar models overestimate by factors of several the neutrino capture rates that have been observed at the earth for different branches of the core fusion reactions, a situation that has not quantitatively improved for more than thirty years (Bahcall et al. 2001).[4]

The expected solar luminosity increase, though difficult to observe directly and so to confirm, should have affected the evolution of the terrestrial environment. Sagan and Mullen (1972) first estimated the terrestrial consequences of an early earth whose surface temperature would apparently be below the freezing point of water until an age of about two Gyr, all other things being equal. Pressed by evidence for liquid oceans generally present on earth as far back as 4.3 Gyr, the well-known "faint-early sun problem" does not presume all things have remained equal. In their excellent summary, Kasting and Catling (2003) emphasize that the rate of silicate weathering of rocks slows as temperature falls, resulting in a slowed removal of carbon dioxide from the air (the time scale for the response is 10^5–10^6 years). That mechanism could provide a slow feedback process that prevented most frozen-ocean states on the ancient earth under its feebler sun by means of a greatly enhanced greenhouse effect.

ESTIMATING THE SUN'S POSSIBLE TERRESTRIAL EFFECTS

Eddy (1976) noted that the low sunspot era of the seventeenth century known as the Maunder Minimum coincided with low solar activity inferred from radiocarbon records and sightings of aurorae, as well as a cold period during the Little Ice Age, then primarily defined for Central and Western Europe (e.g., Lamb 1990). A crude adjustment of then known climate sensitivity led to a hypothetical decrease in total solar irradiance of about 1.4 percent in the Little Ice Age compared to that of the late twentieth century (Eddy 1976, footnote 77). Note that the speculation occurred about a decade prior to satellite measurements of total solar irradiance that indicated a dimmer sun during the Maunder Minimum and other sustained periods of low solar magnetic activity. Such a large irradiance change has been ruled implausible by recent climate simulations and solar theory (e.g., Soon et al. 1996) and also measurements of decadal brightness changes of sunlike stars (Radick et al. 1998).

Modern correlations (figure 9.4) have been found between terrestrial temperature change and solar magnetic variations (Friis-Christensen and Lassen 1991; Baliunas and Soon 1995). Those newer correlations, seen in connection with the sympathetic changes observed in total solar irradiance and solar magnetism, have prompted climate simulation experiments (e.g., Soon et al. 1996; Stott et al. 2003) indicating that solar total irradiance extrapolated back in time explains approximately half the temperature variance of the twentieth century's globally averaged instrumental surface record. In particular, the warming trend of the early twentieth century's globally averaged surface temperature is well simulated with the inclusion of estimated total solar irradiance increases. Similarly, the simulative effect of the air's increased greenhouse gases in the late twentieth century is consistent with the

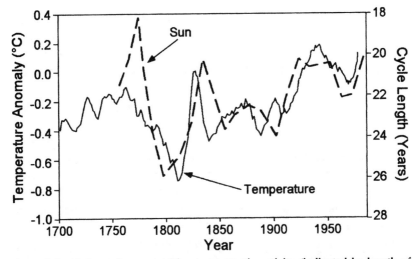

Figure 9.4. Solar surface magnetism or magnetic activity (indicated by length of the twenty-two year or Hale polarity cycle) and Northern Hemisphere temperature record from 1750 to 1984 (see Baliunas and Soon 1995). The curve ends in 1984 to avoid bias in smoothing endpoints beyond the interval of twenty-year smoothing, the time scale of the physical variable (the length of the twenty-two-year cycle) and bias from the inclusion of the period of significant radiative forcing by human-produced greenhouse gases.

surface temperature trend, suggesting a segue to the present period when the contribution by total solar irradiance has been small.

However, several studies have emphasized that total solar irradiance change may not be the sole or dominant solar influence on climate on decade to century time scales (see, e.g., Soon et al. 1996; Soon et al. 2000a, b; Stott et al. 2003). As mentioned previously, solar irradiance integrated over all wavelengths varies no more than 0.15 percent over a sunspot cycle; at the same time the ultraviolet irradiance changes by 10 percent, and with a smaller amplitude, ~1 percent, at wavelengths between 200 and 300 nm. Measurement of the near infrared irradiance recently began with NASA's SORCE (Solar Radiation and Climate Experiment) satellite.

Estimating the sun's ultraviolet interdecadal irradiance is difficult (Fontenela et al. 1999; White et al. 2000; Fox et al. 2004), as is calculating its terrestrial response (Palmer et al. 2004). Suspected terrestrial influences of solar ultraviolet irradiance change include altering the stratospheric ozone concentration, which could affect Hadley cell circulation (Haigh 1996, 1999, 2001) or planetary wave propagation (see Rind 2002). Labitzke and van Loon (1997) mark changes in 30 hPa geopotential heights with the solar cycle; McCormack (2003) simulates photochemical-dynamical changes in the Quasi-Biennial Oscillation, or QBO driven by solar cycle changes in ultraviolet irradiance. An enhanced climate response may arise from solar ultraviolet variability through tropical thick cirrus ice clouds and stratospheric ozone (Soon et al. 2000a). Ogi et al. (2003) suggest that the North Atlantic Oscillation (NAO) affects northern high latitude summer climate over the solar cycle through the intermediary of the cryosphere.

Recall that information obtained from cosmogenic isotope records like ^{10}Be and ^{14}C reveals solar magnetic variability carried by way of cosmic ray by-products. Considering cosmogenic isotope records as exo-terrestrial, rather than narrowly solar—a historical bias perhaps arising from the strong poetry of the sunspot record—leads to additional possible external terrestrial influences. For example, decadal cosmic ray flux variations have been reportedly linked to fractional cloud coverage by low-altitude, watery clouds (Friis-Christensen and Lassen 1991; see further comments by Soon et al. 2000b, Laut 2003, and rebuttals; Kristjánsson et al. 2004). Accompanying increased incident cosmic ray fluxes at phases of low solar magnetism during the decadal cycle is the increased flux of fast moving ions expelled from the sun (sometimes confusingly called "solar cosmic rays") or both, which may impact terrestrial environments (Soon et al. 2000b). Some measurements of cloud cover support the conjecture that at periods of increased flux of cosmic ray and solar ions at the earth (i.e., at sunspot minimum, when solar magnetism is low), low cloud coverage increases by a few percent, thereby lowering the temperature (e.g., Friis-Christensen and Lassen 1991; Lassen and Friis-Christensen 1995); this is also the phase of low total solar irradiance during the decadal cycle. However, the retrieval of cloud properties from different satellite experiments to create a reliable cloud cover record spanning years is still uncertain, as is calculation of the climate impact of the types of clouds that might be created by exo-terrestrial ions, as well as how the ions would form clouds (see discussion in Soon et al. 2000b; Tinsley et al. 2000).

The Microwave Sounding Unit (MSU) satellite record of lower troposphere temperature (Christy et al. 2003) does show a decadal variation that is in phase with the solar coronal hole area, an index of solar particle flux to the earth, phased similarly to the cosmic ray flux to the earth, and antiphased with the sunspot cycle, solar magnetic flux, and total solar irradiance (Soon et al. 2000b). Approximately 50 percent of the observed decadal variance of the lower troposphere temperature record over the twenty-year period studied may stem from the exo-terrestrial influences, possibly particle flux variation. In other words, when the decadal flux of exo-terrestrial ions peaks, the lower troposphere temperature tends to cool; whether this response is mediated by specific cloud properties modulated by exo-terrestrial ions is unknown. Because the phase of peak flux in exo-terrestrial particles closely coincides with total solar irradiance minimum and vice versa, both proposed mechanisms (and any quick-acting amplifying mechanisms) may work in synchrony.

Over the past hundred years the heliospheric magnetic field has been estimated to have increased slowly along with the sun's increased magnetic activity. As a result, the cosmic ray flux has been slowly decreasing (Lockwood et al. 1999), leading to the possibility that cloud formation by exo-terrestrial particles has decreased at the same time, perhaps providing a component of a twentieth-century warming trend, if cloud formation can be so linked to globally averaged surface temperature.

PALEO-ENVIRONMENTAL INDICATORS

Correlations between local environmental parameters and solar variability over the last 1,000 years or longer have recently become numerous, in part because of advances in accelerator mass spectrometry. Stuiver et al. (1997) show variations of the ratio of an oxygen isotope, $\delta^{18}O$, measured in the high-resolution Greenland GISP2 ice core, align with the well-known Oort,

Wolf, Spörer, and Maunder solar activity minima on century scales derived from radiocarbon production rates (as labeled in figure 9.2). Annual layers in sea floor corings from Cariaco Basin off northern Venezuela contain compositional changes marking variations in trade wind intensity and rainfall as the location of the Atlantic Intertropical Convergence Zone (ITCZ) meanders (Haug et al. 2003). In that record, the Little Ice Age, a period linked to low solar activity, is prominent. Corings of the floor of Lake Chichaucarrab in Yucatan reveal a 200-year periodicity in chemical and isotope composition coinciding with radiocarbon production anomalies from sediment in nearby Lake Punta Laguna. Both coring records show anomalies during the Maunder, Spörer, and Wolf solar activity minima (Hodell et al. 2001). Corings from the western continental shelf of the Barents Sea show nonstationary events on century and millennial time scales in the early Holocene that coincide roughly with fluctuations in ^{10}Be production, plus periodicities reminiscent of the sun, namely ~200 years and ~100 years (Sarnthein et al. 2003).

Perturbations in oxygen isotope ratios between 9,000 and 6,000 years ago derived from stalagmites in Oman produced by interdecadal to century shifts in monsoon intensity and tropical rainfall correlate with radiocarbon concentration from tree growth rings (Neff et al. 2001). The oxygen isotope ratio of plant cellulose in peat in northeast China yields a 6,000-year record with variations that correspond "nearly one to one" with those of radiocarbon from tree rings (Hong et al. 2000). In Scandinavia, glacier variations and tree growth limits are correlated with radiocarbon variations, especially for cold events occurring over centuries (Karlén and Kuylenstierna 1996). A sediment record from Rice Lake, North Dakota, implies salinity changes over the past 2,100 years with periodicities ~400, 200, and 100 years, plus lake salinity perturbations corresponding to radiocarbon production anomalies (Yu and Ito 1999). A geographically widespread, sharp cooling ~10,300 years ago and lasting around 200 years coincides with a dramatic increase in cosmogenic isotope production rates and may thus have been driven by solar change (Björk et al. 2001). A sediment record from Arolik Lake, Alaska, spanning the period 12,000 to 2,300 years ago details lake productivity evidenced by swings in diatom populations with periodicities "markedly consistent" with those seen in radiocarbon and ^{10}Be production, such as ~200 years (Hu et al. 2003). Wiles et al. (2004) studied mountain glaciers in Alaska and found expansions consistent with the 200-year solar periodicity, perhaps with amplification mechanisms through the Pacific Decadal Oscillation and Arctic Oscillation.

Millennial-scale periodicities have also been observed in terrestrial reservoirs and linked to inflections in cosmogenic isotope records. Observed ecosystem periodicities of ~1,000 to 1,500 and ~2,400 years have been suggested as originating in solar variability (or more properly, in exo-terrestrial variability, either caused directly by or modulated by the sun). Little guidance can be given from solar models or theory at present, but few solar physicists would be surprised if the sun exhibited such periodicities. Thus, the starting approach has been to search for joint variations, including common periodicities, cross-correlations or specific events. A few examples follow; and one should recall the difficulties of establishing variances of periodicities in cosmogenic and environmental records, as mentioned above (Wunsch 2000, 2003, 2004).

Examining the glacial period from 40,000 to 10,000 years ago, van Geel et al. (1999 and references therein) review evidence for the sharp cold phases of Dansgaard-Oeschger events, with a reported periodicity ~1,500 years in the GISP2 ice core and debris rafted by ice in the subarctic North Atlantic (Bond et al. 2001).[5] Corings from the Santa Lucia Bank, offshore and west of Point Arguello, California, yield ~130,000-year record of Dansgaard-Oeschger cold phases (Kennett et al. 2000). Canadian peat formation during the Holocene shows a ~1,450-year periodicity (Campbell et al. 2000). Masson et al. (2000) find warm events occurring on millennial scales during the Holocene in several Antarctic ice core records; the interval between warm events lengthens (>1,200 years) during cold periods compared to the interval (~800 years) during warm periods. Although Hu et al.'s (2003) record of diatoms from Arolik Lake in Alaska shows a periodocitiy at ~1,500 years, it is very weak in significance; the Arolik Lake productivity record shows a stronger period near ~950 years; the Arolik Lake productivity record does follow inflections in the North Atlantic's record ice-rafted debris.

One notable conclusion is that the moderate amplitudes of local ecosystem change seen during the Holocene, although often weaker than amplitudes observed in the Pleistocene glacial period, indicate variability of ecosystem parameters considerably greater for the Holocene than previously assumed. Also, similar events or periodicities are seen in proxies spread over geographically wide locations, suggesting hemispheric or interhemispheric patterns. That fact is often cited as supporting exo-terrestrial influences to local terrestrial environments.

What of the relation of those millennial-scale environmental variations to the sun's magnetic variability? Results are discrepant, owing at least in part to uncertainties in records and differences in analyses. The GISP2 coring also yields a record of ^{10}Be production, whose inflections van Geel et al. (1999) argue occur near the timings of the ~1,500-year spaced cold events.

However, Stuiver et al. (1997) suggest only a weak solar influence deduced from the radiocarbon record both during the glaciation and early Holocene. Bond et al. (2001) associate millennial cold phases with sustained bundles of century periodicities of high production of cosmogenic nuclide events in both ^{14}C and ^{10}Be records, and suggest a solar forcing.[6] Calibration improvements to the ages should help resolve questions about the existence of exo-terrestrial correlations with ecosystem change on millennial scales.

Explanations of millennial-scale environmental variability are far from complete. If exo-terrestrial are matched by solar activity variations—hence, driven or modulated by the sun—then terrestrial mechanisms to amplify climate responses may be required, as suggested by, for example, Bond et al. (2001) and Rahmstorf (2003)[7]. The presumed exo-terrestrial influences would be expected to have continued into the present and continue into the future—in the absence of an explanation of why they would suddenly cease in the late twentieth century—thus making apportionment of and forecasts for natural factors in climate uncertain in the face of such as yet unknown and unquantified processes.

SUNLIKE STARS

Because solar magnetism appears to be a universal process, other stars with nontrivial subsurface convective zones—by number, approximately 90 percent of the stars in the observable universe—provide enough members that some might stand as examples of the sun at different phases of century to millennial-scale variations (Wilson 1978). Records of surface magnetism in stars (figure 9.5) go back thirty-eight years at Mount Wilson Observatory, beginning with Wilson's (1978) monthly measurements of a sample of nearly one hundred stars, some of which are counterparts to the sun at different ages (Baliunas et al. 1995). Not only surface magnetism but also parallel records of highly precise, differential photometric variations in blue and yellow light have been made in studying the universality of stellar magnetic activity and brightness changes in stars with surface magnetism. The term brightness signifies that the total luminosity has not been observed, but rather more limited spectral information that likely relates to total solar irradiance (Radick et al. 1998). The patterns of variability in lower main sequence stars fall into two broad groups: (1) decadal variations and relatively flat multidecadal periods, similar to the solar decadal variability and Maunder Minimum, found in stars that are older than several billion years, and often called sunlike stars; and (2) high average levels of

Surface magnetic activity in lower main sequence stars

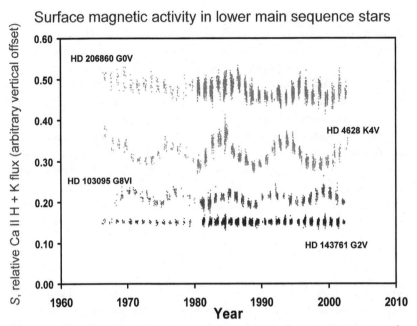

Figure 9.5.   Surface magnetism or magnetic activity in four stars on or near the lower main sequence from a survey begun by O. C. Wilson in 1966 at Mount Wilson Observatory. Surface magnetism is measured as the ratio of the flux in the emission cores of singly-ionized calcium lines in the violet (the Fraunhofer H and K lines at 393.3 and 396.8 nm) and photospheric flux in nearby regions of the spectrum, necessarily integrated over the unresolved stellar disk. The strength of the H and K fluxes increases as the coverage by and intensity of magnetic surface features increases; on the sun the H and K fluxes vary nearly in phase with the sunspot cycle. The four records show the counterpart of the sun approximately 2 billion years ago (top panel, HD 206860), and then three stars, which show records similar in morphology to the record of the present-day sun, HD 4628, HD 103095 and HD 143761. Both HD 4628 and HD 103095 display decadal periodicities, as does the sun at present. The record of HD 143761 may represent the Maunder Minimum phase of low magnetic activity for the sun. The star HD 103095 is an extremely old (approximately 10 billion years) metal deficient subdwarf and is shown as an example of the persistence of decadal magnetic activity cycles in a star of extreme age compared to the sun. The spectral types are listed next to each record's star name. Arbitrary vertical shifts in the average value of the H and K relative fluxes have been applied in order to display the records without overlap; the offsets are 0.0 (HD 143761), 0.02 (HD 103095), 0.09 (HD 4628), and 0.15 (HD 206860).

surface magnetism, often with erratic rather than morphologically smooth cycles, and generally seen in younger, and hence, more rapidly rotating single stars on the lower main sequence.

   For now, without a better theoretical understanding of the stellar dynamo, inferences about solar variability from sunlike stars remain uncertain because a stellar cohort, no matter how carefully selected, may not pertain to solar variability. For example, Lockwood et al. (2004) find that on decadal scales, the sun's brightness variations, compared to its magnetic activity variations, are somewhat—though significantly—smaller than those of its cohort. Even a cohort chosen from a group of stars close in mass to the sun and coeval because they are members of galactic clusters of solar age show considerable magnetic activity variation—more than expected for the sun on century scales (Giampapa et al. 2003). Large amplitudes of brightness variability implied for the sun by solar cohort studies could indicate either that the cohort is inappropriate, or that the sun's irradiance variations have been relatively small in recent decades but could be larger at other times over century or millennial scales.

CONCLUSION

The history of speculation on the sun–climate connection has been littered with promises unfilled, despite correlations, some of which have turned out to be nonstationary or phantom. Exquisite measurements made by satellite-borne instruments in the late twentieth century yield underwhelming amplitudes of total solar irradiance variation accompanying the very dramatic surface magnetic decadal variability and appear to explain little of the instrumental globally averaged surface warming trend of the past twenty to thirty years. However, the accumulation of numerous, new correlations between observed changes in local ecosystems and solar magnetic variability—which may indicate one or any combination of mechanisms such as variable total solar irradiance, particular wavelengths of solar irradiance, for example, the ultraviolet regions, high-speed particles streaming from the sun, or cosmic ray fluxes. If the correlations are found to be sound and physically meaningful, they would require explanation. Some preliminary pathways have been suggested to produce the observed correlations, including cosmic ray or solar particle effects on terrestrial cloud properties, or an amplified response by ocean–air interactions, or by dynamical interactions of the stratospheric ozone layer and widespread wind patterns.

At present the web of correlations, speculation, and hope, is spun with silken strands too tenuous to be held in place while the central force provided by convergence of knots at the core of the web—physical explanation—is missing. However, the recent scientific literature has been rapidly developing information on solar and other exo-terrestrial influences—the varying cosmic ray flux—on environmental parameters. Viewing the sun as redolent with coruscations in magnetic winds, particles and electromagnetic radiation billowing on scales of seconds to millennia and accompanied by changing fluxes of cosmic rays traveling near the speed of light that produce nothing more adverse than quaint auroral displays and cosmogenic isotope blips in records from environmental repositories seems an absurd assumption to hold while facing observed past ecosystem change and their evident correlations with solar variations.

While new empirical links between the sun and climate are being reported, their physical explanations lie in the realm of poorly described, so they cannot be ruled out on scales of decades to millennia. Thus this remnant of the philosophy of solar uniformitarianism—or its contemporary version of benign quasi-solar uniformitarianism—may be pushed to its very verge of toppling, but only after physical processes of the sun and climate are better known.

ACKNOWLEDGMENTS

Willie Soon is gratefully acknowledged for generous insight on this topic. I thank Edouard Bard and Grant Raisbeck for sending their solar activity reconstructions based on cosmogenic data.

NOTES

1. The term "periodicity" is mathematically imprecise and ranks with other terms in the solar literature, for example, pseudo-periodicity, quasi-periodicity or cyclicity; its import is to denote variations that tend to occur on preferred time scales. See Wunsch (2003) for discussion in the context of analysis of past environmental records.

2. However, Damon and Linick (1986) give evidence for stationarity for a 2,400-year periodicity in radiocarbon records, and Charvátová (2000) proposes the periodic motion of the barycenter of the solar system as an inertial driver of activity on that time scale. See also Fairbridge and Shirley (1987) for a discussion of the de Vries cycle and solar inertial motion. Relatedly, Jupiter-mass exo-planets have been discovered orbiting very closely (e.g., nearer than Mercury is to the sun) nearby stars. Decadal magnetic variability has been observed for some of those stars (Baliunas et al. 1995); such exo-planetary systems may allow testing for inertial effects on stellar magnetic activity in a few cases.

3. Hale's vision of the observatory as an unparalleled junction of excellent sky conditions, the largest telescope in the world and brilliant scientific minds led to the

birth of, among many modern astrophysical topics, cosmology. For it was with the splendid 100-inch telescope that Edwin Hubble, along with Milton Humason, defined the enormous spatial scale of the cosmos and its finite temporal existence, having formed some fifteen Gyr ago.

4. The standard models cited here assume nonoscillating neutrinos in electroweak theory.

5. Rahmstorf (2003) suggests a "clock" of 1,470 years with a precision so striking that the periodicity becomes "one of the key issues in climatology that needs to be explained." Rahmstorf notes that the signal is not always present, but is perhaps triggered after the onset of a threshold process, identified as small variations in the freshwater budget of the Nordic Seas. Wunsch (2000) persuasively argues that the sharp spectral peak, almost monotone, is a hazard of the record's undersampled seasonal cycle, which has produced an alias at 1,470 years. Beyond the uncharacteristically sharp peak, the spectrum shows red noise in the millennial region.

6. The supplemental material (supplemental figure 1) to the article describes that the calendar ages of points in the cosmogenic isotope records were shifted to the marine sediment record "to improve the correlations." All adjustments were within two standard deviation errors (approximately 400 years) of the calendar age.

7. This latter paper also suggests that the Little Ice Age may be the most recent of the cold pulses, thereby creating the expectation that the last several centuries are part of a phase of a natural surge toward warmth.

REFERENCES

Bahcall, J. N., M. H. Pinsonneault, and S. Basu. 2001. Solar models: Current epoch and time dependencies, neutrinos, and helioseismological properties. *Ap J* 555:990–1012.

Baliunas, S., and W. Soon. 1995. Are variations in the length of the activity cycle related to changes in brightness in solar-type stars? *Ap J* 450:896–901.

Baliunas, S., et al. 1995. Chromospheric variations in main sequence stars. *Ap J* 438:269–87.

Bard, E., G. M. Raisbek, F. Yiou, and J. Jouzel. 1997. Solar modulation of cosmogenic nuclide production over the last millennium: Comparison between 14C and 10Be records, Earth Planet. *Sci Lett* 150:453–62.

Beer, J. 2000. Long term indirect indices of solar variability. *Space Science Reviews* 94:53–66.

Björk, S., et al. 2001. High-resolution analyses of an early Holocene climate event may imply decreased solar forcing as an important climate trigger. *Geology* 29: 1107–10.

Bond, G., et al. 2001. Persistent solar influence on North Atlantic climate during the Holocene. *Science* 294:2130–36.

Bray, R. J., and R. E. Loughhead. 1964. *Sunspots*. New York: Dover.

Brekke, A., and A. Egeland. 1983. *The Northern Light: From mythology to space research.* Berlin: Springer.

Brougham, Lord. 1803. Critique. *Edinburgh Review.* January.

Campbell, I. D., C. Campbell, Z. Yu, D. H. Vitt, and M. U. Apps. 2000. Millennial-scale rhythms in peatlands in the western interior of Canada and in the global carbon cycle. *Quaternary Res* 54:155–58.

Carrington, R. C. 1860. Description of a singular appearance seen on the Sun on September 1, 1859. *Mon Not R Astron Soc* 20:13.

Charbonneau, P., J. Christensen-Dalsgaard, R. Henning, R. M. Larsen, J. Schou, M. J. Thompson, S. Tomczyk. 1999. Helioseismic constraints on the structure of the solar tachocline. *Ap J* 527:445–60.

Charvátová. I. 2000. Can origin of the 2400-year cycle of solar activity be caused by solar inertial motion? *Ann Geophysicae* 18:399–405.

Chase, T. N., et al. 2001. Relative climatic effects of land cover change and elevated carbon dioxide combined with aerosols: A comparison of model results and observations. *Journ Geophys Res* 106:31, 685–31, 691 no. 2000JD000129.

Chase, T. N., N. Thomas, R. A. Pielke Sr., B. Herman, and X. Zeng. 2004. Likelihood of rapidly increasing surface temperatures unaccompanied by strong warming in the free troposphere. *Climate Res* 25:185–90.

Christy, J. R., R. W. Spencer, W. B. Norris, W. D. Braswell, and D. E. Parker. 2003. Error estimates of Version 5.0 of MSU/AMSU bulk atmospheric temperatures. *J Atmos Oceanic Tech* 20:613–29.

Damon, P. E., and T. W. Linick. 1986. Geomagnetic-heliomagnetic modulation of atmospheric radiocarbon production. *Radiocarbon* 28:266–78.

Damon, P. E., and C. P. Sonett. 1991. Solar and terrestrial components of the atmospheric 14C variation spectrum. In *The sun in time.* Edited by C. P. Sonett, M. S. Giampapa, and M. S. Matthews. Tucson: University of Arizona Press.

DeVorkin, D. H. 1998. Charles Greeley Abbot, biographical memoirs. *Nat Acad Sciences* 73:1–23.

Douglass, D. H., and B. D. Clader. 2002. Climate sensitivity of the Earth to solar irradiance. *Geophys Res Lett* 29:1–4, doi:10.1029/2002GL015345.

Eddy, J. A. 1976. The Maunder minimum. *Science* 192:1189–92.

Fontenela, J. M., Harder J., G. Rottman, T. N. Woods, G. M. Lawrence, and S. Davis. 2004. The signature of solar activity in the infrared spectral irradiance. *Ap J* 605:L85–L88.

Fontenela, J., O. R. White, P. A. Fox, E. H. Avrett, and R.L. Kurucz. 1999. Calculation of solar irradiances. I. Synthesis of the solar spectrum. *Ap J* 518:480–99.

Fairbridge, R. W., and J. H. Shirley. 1987. Prolonged minima and the 179-yr cycle of the solar inertial motion. *Sol Phys* 110:191–220.

Foukal, P. 2002. A comparison of variable solar total and UV irradiances in the 20th century. *Geophys Res Lett* 29:2089–92, doi: 10.1029/2002GL015474.

Fox, P. A., J. M. Fontenela, and O. R. White. 2004. Solar irradiance varibaility: A comparison of models and observations. *Advances in Space Research* 34:231–36.

Frick, P., et al. 1997. Wavelet analysis of solar activity recorded by sunspot groups. *Astron Astrophys* 328:670–81.

Friis-Christensen, E., and L. Lassen. 1991. Length of the solar cycle: An indicator of solar activity closely associated with climate. *Science* 254:698–700.

Fröhlich, C. 2000. Observations of irradiance variations. *Space Science Reviews* 94: 15–24.

Giampapa, M., J. C. Hall, R. R. Radick, and S. L. Baliunas. 2003. Chromospheric activity in solar-type stars. American Astronomical Society Solar Physics Division Meeting 34, abstract 07.10.

Gleissberg, W. 1944. A table of secular variations of the solar cycle. *Terr Magn Atm Electr* 49:243.

Gurnett, D. A., W. S. Kurth, and E. C. Stone. 2003. The Return of the Heliospheric 2–3 kHz Radio Emission During Solar Cycle 23. *Geophys Res Lett* 30(23): 2209, doi:10.1029/2003GL018514.

Haigh, J. 1996. Impact of solar variability on climate. *Science* 272:981–84.

Haigh, J. 1999. A GCM study of climate change in response to the 11-year solar cycle. *Q J Roy Met Soc* 125:871–92.

Haigh, J. 2001. Climate variability and the influence of the sun. *Science* 294:2109–11.

Hale, G. E. 1905. A study of the conditions for solar research at Mount Wilson, Calif. *Ap J* 21:124–50.

Hale, G. E. 1908. On the probable existence of a magnetic field in sun-spots. *Ap J* 28:315–43.

Hale, G. E., and S. B. Nicholson. 1925. Law of sun-spot polarity. *Ap J* 62:270–300.

Haug, G. H., D. Gunther, L. C. Peterson, D. M. Sigman, K. A. Hughen, and B. Aeschlimann. 2003. Climate and collapse of the Mayan civilization. *Science* 299:1731–35.

Herschel, W. 1801. Observations tending to investigate the nature of the Sun, in order to find the causes and symptoms of its variable emission of light and heat. *Phil Trans Roy Soc London* 91:261–333.

Hickey, J. R., B. M. Alton, H. L. Kyle, and D. Hoyt. 1988. Total solar irradiance measurements by ERB/Nimbus-7, a review of nine years. *Space Science Reviews* 48:321–42.

Hodell, D. A., M. Brenner, J. H. Curtis, and T. Guilderson. 2001. Solar forcing of drought frequency in the Maya lowlands. *Science* 292:1367–70.

Hodgson, R. 1860. On a curious appearance seen in the Sun. *Mon Not R Astron Soc* 20:16.

Hong, Y. T., et al. 2000. Response of climate to solar forcing recorded in a 6000-year δ18O time-series of Chinese peat cellulose. *The Holocene* 10:1–7.

Houghton, J. T., et al. (eds.). 2001. *Climate change 2001: The scientific basis, contribution of working group I to the third assessment report of the intergovernmental panel on climate change.* Cambridge: Cambridge University Press.

Hoyt, D. V., and K. H. Schatten. 1997. *The role of the sun in climate change.* Oxford: Oxford University Press.

Hoyt, D. V., and K. H. Schatten. 1998. Group sunspot numbers: A new solar activity reconstruction. *Solar Phys* 181:491–512.

Hu, F. S., et al. 2003. Cyclic variation and solar forcing of Holocene climate in the Alaskan subarctic. *Science* 301:1890–93.

Karlén, W., and J. Kuylenstierna. 1996. On solar forcing of Holocene climate: Evidence from Scandinavia. *The Holocene* 6:359–65.

Kasting, J. F., and D. Catling. 2003. Evolution of a habitable planet. *Ann Rev Astronand Astrophys* 41:429–63.

Kennett, J. P., E. B. Roark, K. G. Cannariato, B. L. Ingram, and R. Tada. 2000. Latest Quaternary paleoclimate and radiocarbon chronology, Hole 1017E, southern California margin. In *Proc. Ocean Drilling Program, Scientific Results,* 167:2 49–254. Edited by M. Lyle, I. Koizumo, C. Richter, and T. C. Moore Jr.

Kristjánsson, J. E., J. Kristiansen, and E. Kaas. 2004. Solar activity, cosmic rays, clouds, and climate: An update. *Adv Space Res* 34:407–15.

Kuhn, J. 2004. Irradiance and solar cycle variability; clues in cycle phase properties. *Adv Space Res* 34:302–7.

Kuhn, J., K. G. Libbrecht, and R. H. Dicke. 1988. The surface temperature of the sun and changes in the solar constant. *Science* 242:908–11.

Kukla, G., and J. Gavin. 2003. Milankovitch climate reinforcements. *Global and Planetary Change* 40:27–48, doi:10.1016/S0921-8181(03)00096-1.

Labitzke, K., and H. van Loon. 1997. The signal of the 11-year sunspot cycle in the upper troposphere–lower stratosphere. *Space Sci Rev* 80:393–410.

Lamb, H. H. 1990. *Climate: Present, past, and future.* New York: Routledge.

Lassen, K., and E. Friis-Christensen. 1995. Variability of the solar cycle length during the past five centuries and the apparent association with terrestrial climate. *J Atmos Solar-Terr Phys* 57:835–45.

Laut, P. 2003. Solar activity and terrestrial climate: An analysis of some purported correlations. *J Atmos Solar-Terr Phys* 65:801–12.

Le, G.-M., and J.-L. Wang. 2003. Wavelet analysis of several important periodic properties in the relative sunspot numbers. *Chin J Astron Astrophys* 5:391–94.

Leighton, R. B. 1969. A magneto-kinematic model of the solar cycle. *Ap J* 156: 1–26.

Lockwood, G. W., R. R. Radick, G. W. Henry, and S. L. Baliunas. 2004. A comparison of solar irradiance variations with those of similar stars. *American Astronomical Society Meeting 204, Abstract 204.0304L.* Contains updates of R. R. Radick, G. W. Lockwood, B. A. Skiff, S. L. Baliunas. 1998. Patterns of variation among sun-like stars. *Ap J Suppl Ser* 118:239–58.

Lockwood, M., R. Stamper, and M. N. Wild. 1999. A doubling of the sun's coronal magnetic field during the past 100 years. *Nature* 399:437–39.

Masson, V., et al. 2000. Holocene climate variability in Antarctica based on 11 ice-core isotopic records. *Quaternary Res* 54:348–58.

Maunder, E. W. 1890. *Monthly Notices of the Royal Astronomical Society* 50:251.

Maunder, E. W. 1894. *Knowledge* 17:173.

McCormack, J. P. 2003. The influence of the 11-year solar cycle on the quasi-biennial oscillation. *Geophys Res Lett* 30(22):2161, doi:10.1029/2003GL018314.

McCracken, K. G., J. Beer, and F. B. McDonald. 2004. Variations in the cosmic radiation, 1890–1986, and the solar and terrestrial implications. *Adv Space Res* 34: 397–406.

Mitchell, J. 1976. Overview of climatic variability and its causal mechanisms. *Quatern Research* 6:481–93.

Moldwin, M. B. 2004. Comment on "The predictability of the magnetosphere and space weather." *EOS (Forum)* 85:15.

Neff, U., S. J. Burns, A. Mangini, M. Mudelsee, D. Fleitmann, and A. Matter. 2001. Strong coherence between solar variability and the monsoon in Oman between 9 and 6 kyr ago. *Nature* 411:290–93.

Ogi, M., K. Yamazaki, and Y. Tachibana. 2003. Solar cycle modulation of the seasonal linkage of the North Atlantic Oscillation (NAO). *Geophys Res Lett* 30:2170, doi:10.1029/2003GL018545.

Ortiz, A., S. K. Solanki, V. Domingo, M. Fligge, and B. Sanahuja. 2002. On the intensity contrast of solar photospheric faculae and network elements. *Astron & Astrophys* 388:1036–47.

Palmer, M. A., L. J. Gray, M. R. Allen, W. A. Norton. 2004. Solar forcing of climate: model results. *Adv Space Res* 34:343–48.

Pap, J. M. 1997. Total solar irradiance variability: A review. In *Past and present variability of the solar-terrestrial system: Measurement, data analysis, and theoretical models.* Edited by G. Cini Castagnoli and A. Provenzle. Amsterdam: IOS Press.

Parker, E. N. 1955a. The formation of sunspots from the solar toroidal field. *Ap J* 121:491–507.

Parker, E. N. 1955b. Hydromagnetic dynamo models. *Ap J* 122:293–314.

Parker, E. N. 1958. Dynamics of the interplanetary gas and magnetic fields. *Ap J* 128:664–76.

Parker, E. N. 1994. *Spontaneous current sheets in magnetic fields.* New York: Oxford University Press.

Petrovay, K. 2000. What makes the sun tick? The origin of the solar cycle. In *The solar cycle and terrestrial climate,* ESA Pub. SPA-463 3–14.

Pielke, R. A., Sr. 2002. Overlooked issues in the U.S. National Climate and IPCC Assessments: An editorial essay. *Clim Change* 52:1–11.

Radick, R., G. W. Lockwood, B. A. Skiff, and S. L. Baliunas. 1998. Patterns of variation among sunlike stars. *Ap J Suppl* 118:239–58.

Rahmstorf, S. 2003. Timing of abrupt climate change: A precise clock. *Geophys Res Lett* 30, 17–1—17-4 doi:10.1029/2003GL017115.

Raisbeck, G. M., and F. Yiou. 2004. Comment on "Millennium scale sunspot number reconstruction: Evidence for an unusually active sun since the 1940s." *Phys Rev Lett* 92, doi:10.1103/PhysRevLett.92.199001.

Raisbeck, G. M., F. Yiou, J. Jouzel, and J. R. Petit. 1990. 10Be and δ2H in polar ice cores as a probe of the solar variability's influence on climate. *Phil Trans R Soc London* A330:463–70.

Ram, M., and M. R. Stolz. 1999. Possible solar influences on the dust profile of the GISP2 ice core from central Greenland. *Geophys Res Lett* 26:1043–46.

Rind, D. 2002. The sun's role in climate variations. *Science* 296:673–77.

Sabine, E. 1852. On periodical laws discoverable in the mean effects of the larger magnetic disturbances. *Phil Trans R Soc London* 142:103.

Sagan, C., and G. Mullen. 1972. Earth and Mars: Evolution of atmospheres and surface temperatures. *Science* 177:52–56.

Sarnthein, M., et al. 2003. Centennial-to-millennial-scale periodicities of Holocene climate and sediment injections off the western Barents shelf, 75oN. *Boreas* 32: 447–61.

Sarton, G. 1947. Early observations of sunspots? *Isis* 37:69–71.

Schawbe, H. 1843. *Astr Nachr* 20(495).

Schwarzschild, M., R. Howard, and R. Härm. 1957. Inhomogeneous stellar models. V. A solar model with convective envelope and inhomogeneous interior. *Astrophysical Journal* 125:233–41.

Shindell, D. T., D. Rind, N. Balachandran, J. Lean, and P. Lonegan. 1999. Solar cycle variability, ozone, and climate. *Science* 284:305–8.

Sonett, C. P., and S. A. Finney. 1990. *Phil Trans Roy Soc London* A330:413.

Soon, W., S. Baliunas, E. Posmentier, and P. Okeke. 2000b. Variations of solar coronal hole area and terrestrial lower tropospheric air temperature from 1979 to mid-1998: Astronomical forcings of change in earth's climate? *New Astronomy* 4:563–79.

Soon, W., E. S. Posmentier, and S. L. Baliunas. 1996. Inference of solar irradiance variability from terrestrial temperature changes, 1880–1993: An astrophysical application of the sun–climate connection. *Ap J* 472:891–902.

Soon, W., E. S. Posmentier, and S. L. Baliunas. 2000a. Climate hypersensitivity to solar forcing? *Ann Geophysicae* 18:583–88.

Soon, W. W.-H., and S. Yaskell. 2004. *The Maunder Minimum and the variable sun-earth connection.* Singapore: World Scientific.

Steenbeck, M., F. Krause, and K.-H. Rädler. 1966. A calculation of the mean electromotive force in an electrically conducting fluid in turbulent motion. *Z Naturforsch* 21a:369–76.

Stott, P. A., G. S. Jones, and J. F. B. Mitchell. 2003. Do models underestimate the solar contribution to recent climate change? *J Climate* 16:4079–93.

Stuiver, M. 1961. *Journal of Geophysical Research* 66:273.

Stuiver, M. 1965. Carbon-14 content of 18th- and 19th-century wood: Variations correlated with sunspot activity. *Science* 149:533–35.

Stuiver, M., and T. F. Braziunas. 1995. Evidence of solar activity variations. In *Climate since A.D. 1500.* Edited by R. S. Bradley and P. D. Jones. New York: Routledge.

Stuiver, M., T. F. Braziunas, B. Becker, and B. Kromer. 1991. Climate, solar, oceanic and geomagnetic influences on late-glacial and Holocene atmospheric 14C/12C change. *Quaternary Res* 35:1–24.

Stuiver, M., T. F. Braziunas, P. M. Grootes, and G. A. Zielinski. 1997. Is there evidence for solar forcing of climate in the GISP2 oxygen isotope record? *Quaternary Res* 48:259–66.

Suess, H. E. 1968. *Meteorolog Monograph* 8:146.

Tinsley, B. A., R. P. Rohrbaugh, M. Hei, and K. V. Beard. 2000. Effects of image charges on the scavenging of aerosol particles by cloud droplets and on droplet charging and possible ice nucleation processes. *J Atmos Sci* 57:2118–34.

Unruh, Y. C., S. K. Solanki, and M. Fligge. 1999. The spectral dependence of facular contrast and solar irradiance variations. *Astron & Astrophys* 345:635–62.

Usoskin, I. G., K. Mursula, S. Solanki, M. Schüssler, and K. Alanko. 2004. Reconstruction of solar activity for the last millennium using 10Be data. *Astron & Astrophys* 413:745–51.

Van Geel, B., O. M. Raspopov, H. Renssen, J. van der Plicht, V. A. Dergachev, and H. A. J. Meijer. 1999. The role of solar forcing upon climate change. *Quaternary Sci Rev* 18:331–38.

Wagner, G., J. Beer, J. Masarik, R. Muscheler, P. W. Kubik, C. Laj, W. Mende, G. M. Raisbeck, and F. Yiou. 2001. Presence of the solar de Vries cycle (205 years) during the last ice age. *Geophys Res Lett* 28:303–6.

Weiss, N. O., and S. M. Tobias. 2000. Physical causes of solar activity. *Space Sci Rev* 94:99–112.

White, O. R., J. Fontenla, and P. A. Fox. 2000. Extreme solar cycle variability in strong lines between 200 and 400 NM. *Space Science Reviews* 94:67–74, doi: 10.1023/A:1026730131648.

Wiles, G. C., R. D. D'Arrigo, R. Villalba, P. E. Calkin, D. J. Barclay. 2004. Century-scale variability and Alaskan temperature change over the past millennium. *Geophys Res Lett* 31:L15203, doi:10.1029/2004GL0200050.

Wilson, O. C. 1978. Chromospheric variations in main-sequence stars. *Ap J* 226:379–96.

Willson, R. C., and A. V. Mordvinov. 2003. Secular total solar irradiance trend during solar cycles 21–23. *Geophys Research Lett* 30(5):1199, doi:10.1029/2002GL016038.

Willson, R. C., and H. S. Hudson. 1988. Solar luminosity variations in solar cycle 21. *Nature* 332:810.

Wunsch, C. 2000. On sharp spectral lines in the climate record and the millennial peak. *Paleoceanography* 15:417–24.

Wunsch, C. 2003. The spectral description of climate change including the 100 ky energy. *Climate Dynamics* 20:353-63, doi:10.1007/s00382-002-0279-z.

Wunsch, C. 2004. Quantitative estimate of the Milankovitch-forced contribution to observed Quaternary climate change, of January 8, 2004, http://puddle .mit.edu/~cwunsch/.

Yu, Z., and E. Ito. 1999. Possible solar forcing of century-scale drought frequency in the northern Great Plains. *Geology* 27:263–66.

10

LIMITATIONS OF COMPUTER PREDICTIONS OF THE EFFECTS OF CARBON DIOXIDE ON GLOBAL CLIMATE

Eric S. Posmentier and Willie Soon

Many scientists forecast disastrous global environmental consequences as a result of projected increases in anthropogenic greenhouse gas emissions. Those researchers base their estimates of those emissions, and the degree of climate change they will cause, on computer climate modeling, a branch of science that has recently made substantial strides in knowledge, but still has significant limitations. The amount of climate change that human activities is expected to cause is relatively small compared with other background and forcing factors (internal and external). Understanding the limits of those models is necessary for any scientific study of modeled climate change.

General circulation model (GCM) calculations share some common deficiencies, whether they are looking at atmospheric temperature, surface temperature, precipitation, or spatial and temporal variability. Those deficiencies arise from complex problems associated with the parameterization of multiple interacting climate components, forcings, and feedbacks, involving especially clouds and the oceans.

Ultimately, we believe that the intrinsic value of a climate model is as a learning rather than predictive tool; in this belief, we echo that of Oreskes et al. (1994). Given the host of uncertainties and unknowns in the difficult but important task of climate modeling, the unique attribution of observed current climate change to increased atmospheric CO_2 concentration, including the relatively well-observed latest twenty years, simply is not possible. We further conclude that the incautious use of GCMs to make future climate projections from incomplete or unknown forcing scenarios is antithetical to the models' inherently instructive value. Unfortunately, such uncritical application of climate models has led to the commonly held but erroneous impression that modeling has "proven" the hypothesis that CO_2 added to the air has caused or will cause significant global warming.

An assessment of the merits of GCMs and their use in suggesting a discernible human influence on global climate can be found in the joint World Meteorological Organization and United Nations Environmental Program's Intergovernmental Panel on Climate Change (IPCC) reports (1990, 1996, and 2001).

Our review highlights only the enormous scientific difficulties facing the calculation of climatic effects of added atmospheric CO_2 in a given GCM. Our purpose is to illuminate areas needing improvement, and to guide policymakers in their understanding of the reliability of model forecasts. Our review neither proves nor disproves a significant anthropogenic influence on global climate. The uncertainty of current estimates of potential human influence on global climate is sufficiently large that they may be greatly overstated, greatly understated, or coincidentally accurate.

CALCULATING THE EFFECTS OF ATMOSPHERIC CO_2

To calculate the effects of increasing atmospheric CO_2 concentration completely and comprehensively, a scientist must overcome three closely related problems:

1. How to calculate the future trajectory of the air's CO_2 concentration.
2. How to calculate its climatic effects.
3. How to separate the CO_2 impacts from other climatic changes.

The first problem involves humanity's impact on the global carbon budget. Anthropogenic emissions of CO_2 are mainly the result of our use of fossil fuels (coal, gas, and oil), which is related to energy consumption and, hence, the world economy. One convenient modeling scheme studies these relationships within the framework of four independent variables: CO_2 released per unit of energy; energy consumed per unit of economic output; economic output per person; and population (Hoffert et al. 1999; Victor 1998). That perspective raises a question: Can economy and technology be sufficiently well prescribed that anyone can reliably predict future energy consumption? And it leads to a subsequent question: What controls the physical exchanges of CO_2 and how do those factors control the apportionment of anthropogenic CO_2 emissions among various reservoirs of the climate system? With respect to these questions, we note that about one third of humanity's carbon production has remained in the atmosphere, with a less certain division between the terrestrial biosphere and oceans (Field and Fung 1999; Joos et al. 1999; Rayner et al. 1999; Giardina and Ryan 2000; Schimel et al. 2000; Valentini et al. 2000; Yang and Wang 2000; Janssens et al. 2003). Of course, economic prediction is a notoriously complex proposition that is even less well defined (Sen 1986; Arthur 1999).

The second and third problems belong strictly to the natural sciences. Here, climate scientists seek a theory capable of describing the thermodynamics, dynamics, chemistry, and biology of earth's atmosphere, land, and oceans. Another fundamental barrier to our understanding and description of the climate system is its inherent unpredictability. Beyond a certain time horizon, even for a deterministic set of conditions, climate simply remains an unknown. (Lighthill 1986; Essex 1991; Tucker 1999). The good news is that attempts to estimate the state of the weather or climate based simply on the so-called primitive equations—the set of fundamental equations governing large-scale atmospheric motions—do yield a finite bound; that is, the results are at least somewhat contained (Lions et al. 1997).

An additional difficulty concerns the logistics of modeling a system with spatial and temporal scales that account for such factors as cloud microphysics and global circulation. Fortunately, that difficulty can be circumvented because of empirical "loopholes" such as the existence of gaps in the energy spectrum of atmospheric and oceanic motions that allow for the separation of various physical and temporal scales. Say, for example, that climate is viewed as an average over a hypothetical ensemble of atmospheric states that are in equilibrium with some slowly changing external factor. Then, under a regular external forcing factor, it seems possible to anticipate the change (Houghton 1991; Palmer 1999).

Essentially all calculations of anthropogenic CO_2 climatic impacts make that implicit assumption (Palmer 1999). But for such a calculation to have predictive value, rather than merely representing the *sensitivity* of a particular model, a GCM must be validated specifically for the purpose of its specific type of prediction. As a case in point, we note that the prediction of climate responses to individual forcings such as the long-lifetime greenhouse gas CO_2, the shorter-lifetime greenhouse gas CH_4, the inhomogenously distributed tropospheric O_3, and atmospheric aerosols would all require separate and independent validation. A logistically feasible validation for such predictions is essentially inconceivable.

The downside of exploiting the energy gap loophole is that relevant physical processes must be parameterized in simple and usable forms. For example, most general circulation models treat radiation with simple empirical schemes instead of solving the equations for radiative energy transfer (Shutts and Green 1978). Chemical and biological changes in the climate system are also highly parameterized. Clearly, some empirical basis and justification for these parameterizations can be made, but because the real atmosphere and ocean have many degrees of freedom as well as connections among processes, there is no guarantee that the package assembled in a general circulation model (GCM) is complete or that it can give us a reliable approximation of reality (Essex 1991).

Going beyond the issue of limited computing resources, Goodman and Marshall (1999) and Liu et al. (1999) have elaborated on various schemes of synchronous and asynchronous coupling for the highly complex atmosphere and ocean GCMs, while warning of the extreme difficulty inherent in deciphering the underlying physical processes of the highly tangled and coupled responses. A call to eschew the direction of all efforts into the scale-resolved physical approach in current formulations of GCMs has also been voiced by Kirk-Davidoff and Lindzen (2000).

Another important point has been raised by Oreskes et al. (1994): It is impossible to achieve a verified and validated numerical climate model because natural systems are never closed, and model results are always nonunique. It follows from Oreskes et al. that the intrinsic value of a climate model is not predictive but heuristic or educational, helping to add to knowledge without providing conclusive fact. The proper use of a climate model is therefore to *challenge* existing formulations (i.e., to test proposed mechanisms of climate change) rather than to *predict* unconstrained scenarios of change such as adding CO_2 to the atmosphere.

SIMULATING CLIMATE VARIABLES

Consider the nominal, globally averaged number of 2.5 Wm^{-2} that is associated with the total radiative forcing provided by the increases of all greenhouse gases (GHGs) since the dawn of the Industrial Revolution. Or, consider a doubling of the air's CO_2 concentration that adds about 3.7 Wm^{-2} to 4.4 Wm^{-2} (IPCC 2001, 356–57) to the troposphere-surface system. To appreciate the difficulties of finding climatic changes associated with these forcings, it is only necessary to consider the energy budget of the entire earth-climate system. Neglecting the nonphysical flux adjustments for freshwater, salinity, and wind stress (momentum) that are also applied in many contemporary GCMs (see discussion in Gordon et al. 2000; Mikola-jewicz and Voss 2000), there are artificial energy or heat flux adjustments as large as 100 Wm^{-2} that are used in some GCMs to minimize unwanted drift in the ocean-atmosphere coupled system (Murphy 1995; Glecker and Weare 1997; Cai and Gordon 1999; Dijkstra and Neelin 1999; Yu and Me-choso 1999).

Models that attempt to avoid artificial heat flux adjustments fare no better because of other substantial biases, including major systematic errors in the computation of sea-surface temperatures and sea ice over many re-gions, as well as large salinity and deep-ocean temperature drifts (Cai and Gordon 1999; Russell and Rind 1999; Yu and Mechoso 1999; Gordon et al. 2000; Russell et al. 2000). Further, the uncertain global energy budgets that are implicit in all GCMs vary by at least 10 Wm^{-2} in empirically de-duced fluxes for shortwave and longwave radiation and latent and sensible heat within the surface-atmosphere system (Kiehl and Trenberth 1997). In addition, Grenier et al. (2000) have called for a simultaneous focus on trop-ical climate drift caused by heat budget imbalances at the top of the atmo-sphere while balancing the surface heat budget because systematic biases in outgoing longwave radiation as large as 10 Wm^{-2} to 20 Wm^{-2} are not un-common in coupled ocean-atmosphere GCMs.

The artificially modified and uncertain energy components of today's GCMs place severe constraints on our ability to find the imprint of a mere 4 Wm^{-2} radiative perturbation associated with anthropogenic CO_2 forcing occurring over 100 to 200 years in the climate system. That difficulty ex-plains why all current GCM studies of the climatic impacts of increased at-mospheric CO_2 are couched in terms of *relative* changes based on control, or unforced, GCM numerical experiments that are already commonly known to be incomplete in their forcing and feedback physics. Soon et al. (1999), for example, identified documented problems associated with mod-els' underestimation or incorrect prediction of natural climate change on decade to century time scales. Some of those problems may be connected to difficulties in modeling both the natural unforced climate variability and suspected climate forcings from volcanic eruptions, stratospheric ozone variations, tropospheric aerosol changes, and variations in the radiant and particle energy outputs of the sun.

Another predicament is the inability of short climatic records to reveal the range of natural variability that would allow confident assessment of probability of climatic changes on time scales of decades to centuries (Soon et al. 2003). Most important, because of our lack of understanding on the range of climatic responses, it is premature to conclude on the basis of the magnitude of forcing alone—4 Wm^{-2} for a doubling of CO_2 vs. 0.4 Wm^{-2} for July insolation changes at 60°N induced by earth's orbital variations over about 100 years, a comparison made by Houghton (1991)—that the climatic changes by human-made CO_2 will overwhelm the more *persistent* effects of a positional change in earth's rotation axis and orbit. The latter form of climate change through gradual insolation change is suspected to be the cause of historical glacial and interglacial climate oscillations, while the potential influence of added CO_2 can only be guessed from our experiences in climate modeling. Indeed, it is a puzzle that there is the clear suggestion of significantly warmer temperatures at both Vostok and Dome Fuji, East Antarctica, during the interglacials at stage 9.3 (about 330,000 years before present; warmer by about 6°C) and 5.5 (about 135,000 years before present; warmer by about 4.5°C) than the most recent one thousand years (Watanabe et al. 2003) despite the relatively low level of atmospheric CO_2 of no more than 300 parts per million (ppm) in the past compared with the present high level of 370 ppm.

Additional historical evidence reveals natural occurrences of large, abrupt climatic changes that are not uncommon (Alley 2000). They occur without any known causal ties to large radiative forcing change. Phase differences between atmospheric CO_2 and proxy temperature in historical records are often unresolved; but atmospheric CO_2 tends to follow rather than lead temperature and biosphere changes (Priem 1997; Dettinger and Ghil 1998; Fischer et al. 1999; Indermühle et al. 1999). In addition, there have been geological times of global cooling with rising CO_2 (during the middle Miocene about 12.5 Myr BP to 14 Myr BP, for example, with a rapid expansion of the East Antarctic Ice Sheet and with a reduction in chemical weathering rates). And there have also been times of global warming with low levels of atmospheric CO_2 (such as during the Miocene Climate Optimum about 14.5 Myr BP to 17 Myr BP, noted by Panagi et al. 1999). To cast anthropogenic or natural CO_2 forcing as the cause of rapid climate change, various complex climatic feedback and amplification mechanisms must operate. But most of those mechanisms for rapid climatic change are neither sufficiently known nor understood (Marotzke 2000; Stocker and Marchal 2000).[1]

Temperature

How well do current GCMs simulate atmospheric temperatures? As Johnson (1997) noted, the appearance of the IPCC (1990) report marks the recognition that all GCMs suffer from the "general coldness problem," par-

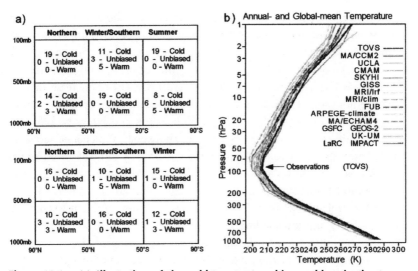

Figure 10.1.   (a) Illustration of the cold-temperature bias problem in the tropo-sphere in simulations produced by fourteen different GCMs. Indicated in each box are the model temperature biases relative to observations (from Johnson 1997). In regions 1, 3, and 5, model results consistently show a cold bias. (b) Note that the cold-bias problem—the fact that most GCM curves lie to the left of the observed temperature line labeled TOVS—extends into the stratosphere (adapted from Pawson et al. 2000).

ticularly in the lower tropical troposphere and upper polar troposphere (regions 1, 3, and 5 in figure 10.1a, which make a total of 105 simulations). Indeed, the general coldness problem was seen in 104 out of the 105 outcomes in regions 1, 3, and 5, from thirty-five different simulations by fourteen climate models.

What is the cause of that ubiquitous error? Johnson (1997) suggests that most GCMs may suffer from extreme sensitivity to systematic physical entropy sources introduced by spurious numerical diffusion, Gibbs oscillations, or inadequacies of sub-grid-scale parameterizations. In other words, the mathematics inherent to the models does not adequately handle all of the processes and potential sources of energy dispersal and transference within the climate system. Even a 4 percent error in modeling net heat flux could result in a biased temperature of 10 degrees, Johnson noted, and net heat flux could be linked to any number of physical sources (including those arising from numerical problems with the transport and change of water substances in forms of vapor, liquid, and ice, and the spurious mixing of moist static energy). The analysis of Egger (1999) seems to support this result and calls for the evaluation of high-order statistical moments (patterns of behavior) to check on the quality of numerical schemes in climate models. By virtue of their strong inherent dependence on temperature, the computation of hydrologic and chemical processes gets contaminated by this critical cold-bias difficulty. In fact, the error in saturation-specific humidity is estimated to double for every 10 degrees of increase in temperature. Johnson (1997) stressed that "erroneous sources of entropy in atmosphere and ocean models differ in both origin and intensity. Efforts in coupled climate modeling to simulate accurately energy exchange across the mutual atmosphere-ocean interface will be extremely difficult. . . . The implication is that atmospheric and oceanic energy balances within coupled climate models that do not require flux adjustment are suspect" (2842).

One probable "solution" to the general coldness problem is simply to add physically unjustifiable flux adjustments for heat (and also for water and sometimes momentum) to the air-sea fluxes. That approach can minimize errors in simulations of the present day climate state, but "may distort their sensitivity to changed radiative forcing" (IPCC 2001). According to IPCC (2001, 479), Marotzke and Stone (1995) show that using flux adjustment to correct surface errors in the control climate does not necessarily correct errors in processes that control the climate change response. On the other hand, models whose parameters have been tuned with "only very loose physical constraints" to simulate today's climate have "larger errors" (IPCC 2001). In short, either approach compromises the reliability of climate change estimates.

The coldness problem also extends to the stratosphere (figure 10.1b), where Pawson et al. (2000) have shown that the cold bias is more uniformly distributed. The range of the cold bias in the global-mean temperatures are about 5–10 degrees in the troposphere and greater than 10 degrees for the stratosphere. Pawson et al. suggest that the particular coldness problem for the stratosphere is more likely associated with problems in physics: For instance, the underestimation of radiative heating rates that results from models' having too little absorption of solar radiation by ozone in the near infrared. Or perhaps there is too much long-wave emission in the middle atmosphere, causing climate models to overcool their stratospheres. Other unresolved problems concern the physical representation of gravity wave momentum deposition in the stratosphere and mesosphere and the generation of gravity waves in the troposphere (McIntyre 1999).

Why discuss the stratosphere when our main concern is the lowest level of the troposphere—the level where plants, animals, and people actually live? There is documented evidence that including this important layer of the atmosphere can improve even *weather* prediction within the troposphere

(Pawson et al. 2000). More important, it has only recently been appreciated that the dynamics of the stratospheric polar vortex, a sort of high altitude jet stream, in close coupling to the vertically propagating patterns of tropospheric flow, is a key parameter governing variability of the troposphere-stratosphere winter circulation under different climate regimes on interdecadal time scales (Kodera et al. 1999; Perlwitz et al. 2000; Tanaka and Tokinaga 2002). Therefore, to address properly the climatic response of added atmospheric CO_2 (or for that matter any number of external forcings under consideration), a GCM must resolve the stratosphere—yet another demanding requirement.

What about surface temperatures? In a systematic comparison of the performance of twenty-three dynamical ocean-atmosphere models, Davey et al. (2002) found that "no single model is consistent with observed behavior in the tropical ocean regions . . . as the model biases are large and gross errors are readily apparent" (418). Without flux adjustment, most models produced annual mean equatorial sea-surface temperature (SST) in the central Pacific that are too cold by 2°C to 3°C. All GCMs except one simulated the wrong sign of the east-west SST gradient in the equatorial Atlantic. The GCMs also incorrectly simulate the seasonal climatology in all ocean sectors and its interannual variability in the Pacific Ocean; surface wind stress is diagnosed as the key parameter leading to those poor outcomes. The shortfall in interannual variability is more pronounced for zonal wind stress than for SST. Schneider (2002) made the first progress in isolating and understanding specific intramodel and intermodel disagreements in the simulations of the equatorial Pacific Ocean climatology and variability by using various flux-corrected experiments.

Also notable is the evaluation by Bell et al. (2000) of the interannual changes in surface temperature of the control (unforced) experiments from sixteen different coupled ocean-atmosphere GCMs of the Coupled Model Intercomparison Project (CMIP) (figure 10.2). Bell et al. found that the majority of the GCMs significantly underestimate the observed, de-trended worldwide averaged surface temperature variability over the oceans (figure 10.2b) while they overestimate such variability over land (figure 10.2c). That systematic difference is most clearly illustrated by the ratio of the over-land to over-ocean temperature variability in panel d of figure 10.2. The authors discuss various factors, such as forcing agents (CO_2, solar variability, and volcanic eruptions) and the GCMs underestimation of El Niño–Southern Oscillation (ENSO) variability, factors that could be responsible for the systematic discrepancy between observed and GCM-predicted interannual temperature variability. They eventually settled on

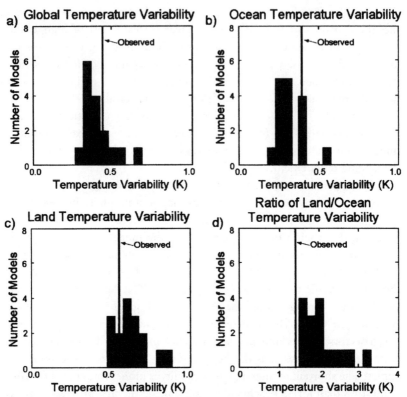

Figure 10.2.   Comparisons of detrended 1959–1998 observed surface temperature variability with the unforced results from sixteen different GCMs of the CMIP (temperature variability is calculated from the r.m.s. standard deviation of the annually averaged data). The statistically significant difference between the observed and the GCM ratios of the land/ocean variability (panel d) has been shown to be associated with an inadequate or incorrect parameterization of land surface processes (adapted from Bell et al. 2000).

nonphysical representations of land surfaces that lead to lower soil moisture and larger land temperature variability than do more realistic land surface schemes. Bell et al. also point out another problem of most GCMs: too much variability in the models' surface temperatures over both land and sea at high latitudes, where excessive interannual variability in the GCMs' predictions of snow and sea ice coverage is also noted.

The findings of Bell et al. (2000) should not be surprising, as physical modeling of land processes is particularly difficult, laden as it is with many unknown factors and large uncertainties. For example, Pitman et al. (1999) determined that for tropical forest, annually averaged simulations varied by

79 Wm^{-2} for the sensible heat flux and 80 Wm^{-2} for the latent heat flux in sixteen different GCMs. Over grassland, the range was 34 Wm^{-2} and 27 Wm^{-2}, respectively. The models' simulations of temperature differed by 1.4 K for tropical forest and 2.2 K for grassland.

Yet another important concern arises from the tradeoff between realism and complexity. For example, new climate drifts appear in atmospheric GCMs with explicit treatment of land variables such as soil moisture or snow water mass (Dirmeyer 2001). Such a serious investment in model complexity is important for numerical weather prediction and may be needed for treating climate forcing by anthropogenic CO_2 (as discussed in section 4).

Precipitation

Soden (2000) has documented a problem in the current generation of GCMs that stems from the inability of some thirty different atmospheric GCMs in the Atmospheric Modeling Intercomparison Project (AMIP) to reproduce faithfully interannual changes in precipitation over the Tropics (30N to 30S). Figure 10.3 depicts the good agreement between observations and the GCMs' simulations of atmospheric water vapor content, tropospheric temperature at 200 mb, and outgoing longwave radiation (OLR), but it also reveals the *poor* agreement between observations and model simulations of precipitation and net downward longwave radiation at the surface. Considering especially the more direct association between the latent heat released during precipitation and the warming and cooling of the atmosphere, Soden (2000) warned that the good agreement between the observed and modeled temperature at 200 mb (figure 10.3c) is surprising in light of the large differences for a simultaneous comparison of the precipitation field (figure 10.3a).

That comparison suggests that the temperature agreement at 200 mb could be fortuitous, since the atmospheric GCMs were forced with *observed* sea-surface temperatures, while the modeled interannual variabilities of the hydrologic cycle are seriously underestimated by about a factor of three. Based on the models' relatively constant values of downward longwave radiation reaching the surface (figure 10.3e; see Wild et al. 2001 for further quantitative comparison around the globe), Soden (2000) points to possible systematic errors in current GCM representations of low-lying boundary layer clouds. However, the study cannot exclude the possibility of errors in algorithms that retrieve precipitation data from observations made by satellites, which would emphasize the need for improved precipitation products.

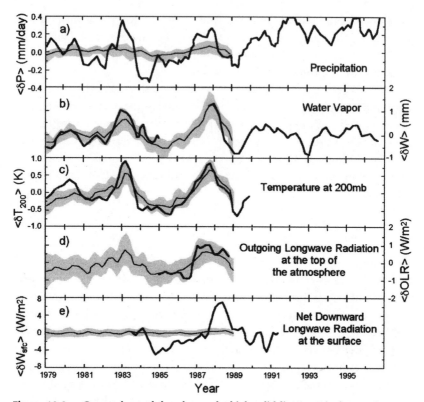

Figure 10.3.   Comparison of the observed (thick solid line) tropical-mean interannual variations of (a) precipitation, (b) total precipitable water, (c) temperature at 200 mb, (d) outgoing longwave radiation (OLR) at the top of the atmosphere, and (e) the net downward longwave radiation at the surface with the ensemble-mean of thirty AMIP GCM results (the thin solid curve overlaid contained in gray region showing the range of one intermodel standard deviation of the ensemble mean). Contrast the good agreement for simulated water vapor, temperature, and OLR with the internally inconsistent results for precipitation and net surface longwave radiation. (All climate simulations were forced with observed sea surface temperatures.) (Adapted from Soden 2000)

Water Vapor

Soden (2000) highlighted the positive ability of GCMs to simulate the correct sign and magnitude of the observed water vapor change in figure 10.3b. This conclusion agrees with the extensive review by Held and Soden (2000) on water vapor feedbacks in GCMs. Held and Soden called for a clearer recognition of GCM's proficiency in calculating the water vapor feedback (which diagnoses model ability to simulate the *residual* between evaporation and precipitation rather than evaporation or precipitation per se) versus GCMs representation of the more complicated physics related to the cloud forcing and feedback.

It is critical, however, to add that the latest analyses of the interannual correlation between tropical mean water vapor content of the atmosphere and its surface value continue to show significant differences for the vertical patterns derived from rawinsonde data and outputs of GCMs, including those of the newer AMIP2 study (Sun et al. 2000). Essentially, in comparison with rawinsonde data, GCMs exhibit too strong a coupling between mid- to upper tropospheric water vapor and surface water vapor. Water vapor in GCMs has also been found to have a stronger dependence on atmospheric temperature than the empirical relation deduced from observations.

Finally, purely numerical problems also exist; they are associated with physically impossible, negative specific humidity in the Northern Hemisphere extratropics caused by problematic parameterization of steep topographical features (Rasch and Williamson 1990; Schneider et al. 1999).

Clouds

In figure 10.4, we show the sensitivity of the parameterization of the large-scale formation of cloud cover that is used in one state-of-the-art model (Yang et al. 2000). As parameterized, cloud cover is extremely sensitive to relative humidity (U in figure 10.4) and to both the saturated relative humidity within the cloud (U_s), and the threshold relative humidity at which condensation begins (U_{00}). The creators of this GCM discuss how the formula is used to tune the formation of clouds (through large-scale condensation at high latitudes or near polar regions) by 20 percent to 30 percent in order to match what is observed.

Wielicki et al. (2002) offered observational evidence for large decadal variability of the tropical mean radiative energy budget of the past two decades, which may be explained by independently observed changes in tropical mean cloudiness. More significantly, those results highlighted "the critical need to improve cloud modeling" because several GCMs failed to simulate that large observed variability in the energy budget tropical ocean-atmosphere system.

Other researchers (e.g., Grabowski 2000) emphasize the importance of the proper evaluation of the effects of cloud microphysics on tropical climate by using models that directly resolve mesoscale dynamics. Grabowski points out that the main effect of cloud microphysics is on the

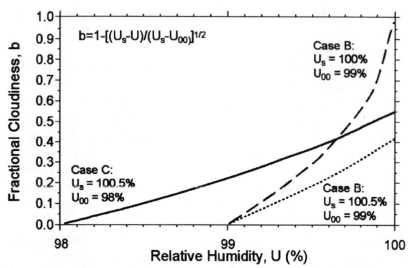

Figure 10.4. The parameterized cloud cover is very sensitive (contrasted by cases A, B, and C) to relative humidity, U, and to values of U_s, the saturated relative humidity within the cloud and U_{00}, the threshold relative humidity at which condensation begins. (Adapted from Yang et al. 2000)

ocean surface rather than directly on atmospheric processes. Because of the great mismatch between the time scales of oceanic and atmospheric dynamics, Grabowski was pessimistic about quantifying the relation between cloud microphysics and tropical climate. Clearly, the parameterizations of cloud microphysics and cloud formation processes, as well as their interactions with other variables of the ocean and atmosphere, remain major challenges for climate modelers.

EXPECTED OUTCOMES OF CO_2 FORCING

Given the range of uncertainties and numerous unknowns associated with parameterizations of important climatic processes and variables, what should we expect from current GCMs for a scenario with an increased CO_2 forcing? The most common difficulty facing the interpretation of many GCMs results is related to confusion arising from imposed natural and anthropogenic forcings that may or may not be internally consistent. This is why Bengtsson et al. (1999) and Covey (2000) have called for more inclusive consideration of all climate forcings, accurately known or otherwise, rather than a piecemeal approach that yields oversimplifications. The enormity and urgency of the scientific task, with modern societal needs for coping with climate change regardless of the added concerns about anthropogenic CO_2 forcing, has led Pielke (2001) to remind us that "there have been no model experiments to assess climate prediction in which all important atmosphere-ocean-land surface processes were included."

Many qualitative outcomes of forcing by anthropogenic greenhouse gases have been postulated: Changes in standard ocean-atmosphere variables of wind, water vapor, rain, snow, land and sea ice, sea-level, the frequency and intensity of extreme events like storms and hurricanes (Soon et al. 1999), as well as more exotic phenomena, including large cooling of the mesosphere and thermosphere (Akmaev and Fomichev 2000), increased presence or brightness of noctilucent clouds near the polar summer mesopause (Thomas 1996; but see Gadsden 1998), increases in atmospheric angular momentum and length of day (Abarca del Río 1999; Huang et al. 2001), and shrinking of surfaces of constant density at operating satellite altitudes (Keating et al. 2000). In these calculations, the benchmark forcing scenario is usually an emission rate of 1 percent per year, chosen to represent roughly the CO_2 equivalent of the burden of all anthropogenic greenhouse gases.

Although some of these studies claim that observations are consistent with modeled CO_2 effects, it is clear that even the theoretical claims, with their strong bias toward accounting for only the effects of GHGs, are neither robust nor internally consistent. A good example is the predicted change of the Arctic Oscillation (AO) pattern of atmospheric circulation by the year 2100. The AO is one of the key variability patterns of the wintertime atmospheric circulation over the Northern Hemisphere, characterized broadly by a redistribution of air mass between polar regions and mid-latitudes. Here, Zorita and González-Rouco (2000) found, using results from two different GCMs and a total of six simulations with different initial conditions, that both upward *and* downward tendencies in the intensity of the AO circulation pattern are likely under the same scenario of increasing atmospheric CO_2. Apparently, internal model variability dominates those effects of the external forcing by CO_2 and leads to an ambiguous expectation for a CO_2-related signal in the modeled AO variability. This reemphasis on unforced internal variability is consistent with the recent classification of the observed vertical structures of the AO into distinct perturbations originating in the troposphere versus stratosphere by Kodera and Kuroda (2000). Besides cautioning about the lack of robustness of previous claims for the AO owing to increased CO_2 forcing, Zorita and González-Rouco highlighted the direct impact of that unknown on the calculation of the Northern Hemisphere's regional climate change in the extratropics.

Some theoretically predicted CO_2 effects are not detectable unless a very high, or even extreme, level of CO_2 loading is imposed. It is also predicted that a transient GCM experiment forced with the slightly lower CO_2 emission growth rate of 0.25 percent per year, as opposed to the present growth rate of 0.4 percent per year, will ultimately lead to a relatively larger sea-level rise (based only on the thermal expansion of sea water; Stouffer and Manabe 1999). An additional 15 cm rise (the calculated global sea level rise for the emission case of 0.4 percent per year is roughly 27 cm) by the time of doubling the atmosphere's carbon dioxide content is expected because the atmospheric heating anomaly of the slower-carbon-dioxide emission-rate world will have more time to penetrate deep into the ocean, thereby causing a relatively larger thermal expansion of seawater and hence a larger rise in the sea level.

One example of a problem with estimating the effects of a high level of atmospheric CO_2 loading concerns potential changes in ENSO characteristics, for which no statistically significant change is predicted until the anthropogenic forcing is four times the preindustrial value (Collins 2000a). On the other hand, Collins (2000b) subsequently reported a surprising result—no significant change in ENSO characteristics occurred for a similar 4xCO_2 numerical experiment, based on an updated GCM with improved horizontal ocean resolution and no heat flux adjustment. Collins concluded that calculating ENSO response to increasing greenhouse-gas forcing can depend sensitively and nonlinearly on subtle changes in model representations of sub-grid processes (rather than depending on the gross model parameters such as ocean resolution and heat flux adjustment that are the main differences between the new and old versions of GCM he used). Thus, Collins concludes, improved understanding of climate change requires further exploration of the parameter-space of coupled ocean-atmosphere GCMs.

As for the statistics of recent ENSO variability, Timmermann (1999) has shown that the observed changes are not inconsistent with the null hypothesis of natural variability of a nonstationary climatic system. In addition, the careful case study by Landsea and Knaff (2000) confirmed that no current climate model provided both useful and skillful forecasts of the entire 1997–1998 El Niño event. The IPCC (2001) did mention that "some aspects of model simulations of ENSO, monsoons, and the NAO, as well as selected periods of past climate, have improved." But the fact remains that the ENSO phenomenon has resisted reliable forecasts by *any* current model (Fedorov et al. 2003). The physics missing from these models is likely to have an equally deleterious effect on forecasts of CO_2-induced climate change.

Expected Changes in Seasonal Temperatures?

We will consider three responses under the typical equivalent CO_2-forcing scenario of 1 percent per year, starting with the seasons. Is the CO_2-forced change expected to alter the character of seasonal cycles? If so, how do predictions compare with what is observed, at least over the last few decades?

Jain et al. (1999) examined this question by considering three parameters for the Northern Hemisphere (NH) surface temperature: the mean temperature's amplitude and phase, the equator-to-pole surface temperature gradient (EPG), and the ocean-land surface temperature contrast (OLC).

A comparison of observed and modeled EPG and OLC climatologies is summarized in table 10.1. The results show that expected changes owing to CO_2 forcing are often very small when compared with differences between the unforced GCM and observed values in EPG and OLC. Hence, detecting CO_2 effects in seasonal differences of EPG and OLC may not be feasible.

Jain et al. (1999) did find significant differences between observed interannual and decadal trends of both EPG and OLC and results obtained from CO_2-forced climate experiments. For example, the CO_2-forced run produced a statistically significant increase in amplitude (and delay in phase) for the seasonal cycle of OLC. But no change was apparent in observed temperatures. Worse yet, even the unforced experiment yielded a statistically significant increase in the amplitude of the OLC seasonal cycle, which makes the search for a CO_2 signal via this means almost impossible. It was determined, however, that the amplitude of the annual cycle of NH surface temperature decreased in a way consistent with results obtained from the CO_2-forced experiment. On the other hand, the observed trend in *phase* shows an *advance* of the seasons rather than the *delay* derived from the

Table 10.1. Observations and predictions (both unforced GCM and CO_2-forced GCM results) of seasonal and annual Northern Hemisphere (NH) equator-to-pole surface temperature gradients (in °C per 5° latitude; EPG) and ocean-land surface temperature contrasts (in °C; OLC). (Jain et al. 1999)

| | EPG | | | OLC | | |
	Annual	Summer (JJA)	Winter (DJF)	Annual	Summer (JJA)	Winter (DJF)
NH Observations	-3.1	-2.0	-3.9	0.3	-5.5	6.5
GCM unforced	-2.9	-1.7	-3.8	3.8	-3.8	11.4
GCM CO_2-forced	-2.7	-1.6	-3.6	3.3	-4.4	10.9

models. Jain et al. offer three possible reasons for the disagreement: the use of model flux corrections, the significant impact of low-frequency natural variability, and sampling problems associated with the observations. An obvious fourth possibility is that the model results are incorrect, and the obvious fifth is that CO_2 forcing has not affected those variables.

In light of these confusions and difficulties, seasonal cycles are probably not good "fingerprints" for identifying the impact of anthropogenic CO_2. This conclusion seems consistent with the independent finding by Covey et al. (2000) that showed seasonal cycle amplitude to depend only weakly on equilibrium climate sensitivity (i.e., equivalent to a varying climate forcing in the present comparison), based on the range of results from seventeen coupled ocean-atmosphere GCMs from the Coupled Model Intercomparison Project. If these results are correct, then it is odd that seasonality in forcing (from geometrical changes in solar insolation by changing tilt angle of earth's rotation axis and earth's orbital position around the sun) is believed to cause very large changes in mean climate, but significant changes in mean forcing; that is, from atmospheric CO_2, cause only insignificant changes in the seasonal climatology. Such an empirical constraint continues to challenge the conceptual framework, especially when assessing the role of added anthropogenic CO_2 and other gases and particulates to the air, of radiative forcing-response of the climate system (see, e.g., page 8 of the Summary for Policymakers of IPCC 2001).

Expected Changes in Clouds?

Next, consider clouds. Given the complexity of representing their relevant processes, can we expect to find a CO_2-forced imprint in clouds?

First, as Yao and Del Genio (1999) have noted, it is misleading to assert that an increased cloud cover is evidence of CO_2-produced global warming (i.e., a warming climate with more evaporation and, hence, more clouds). That is so because cloud cover depends more directly on relative humidity than on specific humidity. For example, under CO_2 doubling experiments with different parameterization schemes, Yao and Del Genio (1999) predicted a *decrease* in global cloud cover, although there was an increase in middle- and high-latitude continental cloudiness. They also cautioned that because "physical basis for parameterizing cloud cover does not yet exist," all predictions about cloud changes in response to rising atmospheric CO_2 concentrations should be viewed carefully.

Others, like Senior (1999), have emphasized the importance of including parameterizations of interactive cloud radiative properties in GCMs and called for a common diagnostic output like the water path length within the cloud in control (unforced) experiments. On another research front, Rotstayn (1999) implemented the detailed microphysical processes of a prognostic cloud scheme in a GCM and found a large difference in the climate sensitivity between that experiment and one with a diagnostic treatment of clouds. A stronger water vapor feedback was noted in the run with the prognostic cloud scheme than in the run with the diagnostic scheme, and the stronger water vapor feedback caused a strong upward shift of the tropopause upon warming. Rotstayn found that an artificial restriction on the maximum heights of high clouds in the diagnostic scheme largely explained the differences in climatic response.

At this stage of incremental learning we conclude that no reliable predictions currently exist for the response of clouds to increased atmospheric CO_2. So sensitive are certain cloud feedbacks to cloud microphysics, for example, that a lowering of the radius of low-level stratus-cloud droplet size from 10 μm to 8 μm would be sufficient to balance the warming from a doubling of the air's CO_2 concentration. Likewise, a 4 percent increase in the area of stratus clouds over the globe could also potentially compensate for the estimated warming of a doubled atmospheric CO_2 concentration (Miles et al. 2000).

Expected Changes in the Oceans?

Finally, consider the oceans. Under an increased atmospheric CO_2 forcing of, say, 1 percent per year, one commonly predicted transient response is a weakening of the North Atlantic thermohaline circulation (THC), owing to an increase in freshwater influx (Dixon et al. 1999; Rahmstorf and Ganopolski 1999; Russell and Rind 1999; Wood et al. 1999; Mikolajewicz and Voss 2000; see figure 10.5a). However, with an improved representation of air-sea interactions in the tropics, the significant weakening (or even collapse under stronger and persistent forcing) of the THC predicted by earlier GCMs cannot be reproduced (Latif et al. 2000; see figure 10.5b). (While considering Latif et al.'s results in figure 10.5b, it is useful to note from figure 10.5a that the coarser version of the MPI model actually did predict a weakening of thermohaline circulation just like other models of figure10.5a.)

In another GCM experiment, Russell and Rind (1999) observed that despite a global warming of 1.4°C near the time of CO_2 doubling, large regional cooling of up to 4°C occurred in both the North Atlantic Ocean (56-80°N, 35°W-45°E) and South Pacific (near the Ross Sea, 60-72°S,

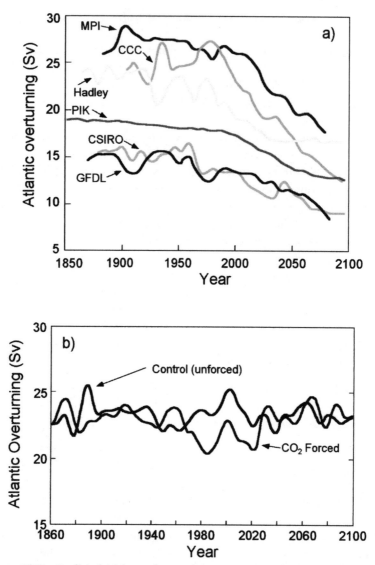

Figure 10.5. Predicted (a) large changes (20–50 percent reduction in overturning rate by 2100 A.D.) in the thermohaline circulation (THC) for six different coupled climate models (adapted from Rahmstorf 1999) vs. (b) a relatively stable THC response in a state-of-the-art MPI GCM with improved spatial resolution of the tropical ocean (adapted from Latif et al. 2000) under a similar CO_2-forced scenario. The quantity shown is the maximum North Atlantic overturning flowrate in Svedrups (10^6 m^3 s^{-1}) at a depth of about 2000 m. Wood et al. (1999) noted, however, that the measure of the THC strength for the meridional overturning adopted here cannot be estimated from observations. They proposed the Greenland-Iceland-Scotland ridge, south of Cape Farewell at the southern tip of Greenland and the transatlantic section at 24N, as three locations where more robust observations are available for comparison with GCM results.

165°E–115°W) because of reduced meridional poleward heat transfer over the North Atlantic and local convection over the South Pacific.

But Russell et al. (2000) later demonstrated that the predicted regional changes over the Southern Ocean were unreliable because of the model's excessive sea ice variability. Another GCM's high-latitude southern ocean suffered a large drift (Cai and Gordon 1999). For example, within 100 years after coupling the atmosphere to the ocean, the Antarctic Circumpolar Current was noted to intensify by 30 Sv (from 157 to 187 Sv, where 1 Sverdrup or Sv equals 10^6 m^3/s), despite the use of flux adjustments. Cai and Gordon identified the instability of convection patterns in the Southern Ocean to be the primary cause of this drift problem.

Mikolajewicz and Voss (2000) further caution that there is still significant confusion about what mechanisms are responsible for the weakening of the THC in various models, since different GCMs give contrasting roles to individual atmospheric and oceanic fluxes of heat, moisture, salinity, and momentum. In addition, several oceanographers (Bryden 1999; Holloway and Saenko 1999) have expressed concern about the lack of both physical understanding and realistic representation of ocean circulation in global models. Criticisms were especially directed toward the highly schematic representation of the North Atlantic THC as a conveyor belt providing linkages to the world's oceans. Holloway and Saenko (1999) state that:

> Understanding what makes the conveyor work is deficient, drawing mainly on the role of buoyancy loss leading to sinking [is] somewhat like trying to push a string. The missing dynamics are that eddies in the presence of bottom topography tend to set up mean flows that carry major circuits of the conveyors, allowing sunken water masses to "go for the ride." Climate models have difficulty in both these regards—to include (if at all!) a plausible Arctic Ocean and to deal with eddies either explicitly or by parameterization.

In spite of the problems in modeling the THC's physical processes, a complete breakdown of the North Atlantic THC is predicted under a sufficiently strong CO_2 forcing (Broecker 1987; Schmittner and Stocker 1999; Rahmstorf 2000; see, e.g., Manabe and Stouffer 1993 for scenarios forced by a quadrupling of atmospheric CO_2). Yet, as Rahmstorf and Ganopolski (1999), Wood et al. (1999), and Mikolajewicz and Voss (2000) point out, the predicted changes of the THC are very sensitive to parameterizations of various components of the hydrologic cycle, including precipitation, evaporation, and river runoff. Hence, without a perpetually enhanced influx of freshwater (from any source) or extreme CO_2 forcing, the transient decrease in THC overturning eventually recovers as time progresses in the model (Holland et al. 2000; Mikolajewicz and Voss 2000). In addition, by including a dynamic sea ice module in a coupled atmosphere-ocean model, Holland et al. (2000) report a reduction (rather than an enlargement) in the variance of the THC overturning flow rate, under the doubled CO_2 condition, down to 0.25 Sv^2 (or only 7 percent) from the high value of 3.6 Sv^2 simulated under the present-day forcing level.

Furthermore, Latif et al. (2000) have just reported a new stabilization mechanism that seems to lower previous expectations of a CO_2-induced THC weakening (figure 10.5b but see also Rahmstorf 2000). In the case of Latif et al., the state-of-the-art coupled ocean-atmosphere GCM of the Max Planck Institut für Meteorologie at Hamburg (MPI) resolves the tropical oceans at a meridional scale of 0.5°, rather than the more typical scale of 2° to 6°, and produces no weakening of the THC when forced by increasing CO_2. Latif et al. showed that anomalously high salinities in the tropical Atlantic (produced by excess freshening in the equatorial Pacific) were advected poleward to the sinking region of the THC; and the effect was sufficient to compensate for the local increase in freshwater influx there. Updated experiments for a similar CO_2 forcing scenario by Gent (2001) and Sun and Bleck (2001) also confirm the relative stability of the THC because of compensating effects between thermal perturbation and changes in surface hydrology and salinity.

Hence, with the additional stabilizing degree of freedom from the tropical oceans, the THC remains stable under that CO_2-forced experiment, leaving no reliable prediction for change in oceanic circulation in the North Atlantic under an added CO_2 climate. Latif et al. (2000) concluded that the response of THC to enhanced greenhouse warming is still an open question. Delworth and Dixon (2000) added another mechanism that could serve to oppose the THC weakening effect in numerical experiments with increasing CO_2. These authors, using their relatively coarser-resolution GCM, found that given an enhanced forcing owing to an increase in the westerly wind speed over the North Atlantic (as inferred from the observed pattern of the Arctic Oscillation over the past thirty years) the THC weakening trend from greenhouse warming scenario could be delayed by several decades. Apparently, the stronger winds over the North Atlantic extract more heat from the ocean and hence cool the upper ocean, and increase its density sufficiently to temporarily counteract some of the effects from net freshening over the North Atlantic. Delworth and Dixon noted, however, that the excess freshening over the North Atlantic does eventually lead to a significant reduction of the THC.

Rahmstorf (2000) summarized all earlier numerical experiments that proposed a significant (20–50 percent) reduction in the THC overturning rate under global warming scenarios by A.D. 2100. We emphasize that our highlighting of the contrasting GCM results by Latif et al. or by Delworth and Dixon, noting the preferable higher spatial resolution of Latif et al.'s GCM, does not undermine all previous models results. The exercise conducted here is meant to note the inconsistency among GCMs for the predicted changes in THC. We conclude that no robust or quantitative prediction of THC is currently possible. Gent (2001) cautioned that estimating the response of the THC in the twenty-first century, important as the question may be, is "a very demanding question to ask of current state-of-the-art coupled climate models."

DEALING WITH THE ISSUES

Many questions remain open concerning what can be deduced from the current generation of GCMs about potential CO_2-induced modifications of earth's climate. The climatic impacts of increases in atmospheric CO_2 are not known with a practical or measurable degree of certainty. Specific attempts to fingerprint CO_2 forcing by comparing observed and modeled changes in the vertical temperature profiles have yielded new insights related to areas where model physics may be improved. One good example is the unrealistically coherent coupling between the lapse rate and tropospheric mean temperature in the tropics for variability over time scales of three to ten years (Gillett et al. 2000).

But even the range of modeled global warming remains large and is not well constrained (Forest et al. 2000). For example, the aggregate of various GCMs projects a global climate warming that ranges from 1.4°C to 5.8°C (IPCC 2001) by 2100 for a range of prescribed climate sensitivity values (about 1.7°C to 4.2°C for equilibrium temperature change from a doubling of atmospheric CO_2) and emission scenarios (thirty-five of those so-called no climate policy *Special Report on Emissions Scenarios* [SRES] based on four "storylines").[2] This range of uncertainty is about 70 percent larger than the earlier projected global warming range of 0.9°C to 3.5°C (IPCC 1996), reflecting perhaps an increasing acceptance of the uncertainties intrinsic to today's state-of-the-art numerical climate models.[3]

Räisänen (1999) more optimistically suggested that many of the qualitative intermodel disagreements in CO_2-forced climate responses (including differing signs of predicted response in some variables—sea level pressure, precipitation, and soil moisture) could be attributed largely to differences in internal variability in different climate models. On the other hand, Räisänen (1999) cautioned that it may be dangerous to rely on a single GCM for the study of climate change scenarios because "a good control climate might partly result from skillful tuning rather than from a proper representation of the feedbacks that are important for the simulation of climate change."

Building partly on that idea, Forest et al. (2000) used the MIT statistical-dynamical climate model to quantify the probability of expected outcomes by performing a large number of sensitivity runs—by varying the cloud feedback and the rate of heat uptake by the deep ocean. It turned out that the IPCC's range of equilibrium climate sensitivity of 1.5°C to 4.5°C corresponded roughly to only an 80 percent confidence interval of possible responses under a particular optimal value of global-mean vertical thermal diffusivity below the ocean's mixed layer. The 95 percent probability range for the climate sensitivity as quantified by Forest et al. was 0.7°C to 5.1°C; and, in the final analysis, Forest et al. (2000) determined the more relevant result for transient responses to a doubling of atmospheric CO_2 to be a mean global warming of 0.5°C and 3.3°C at the 95 percent confidence level. Forest et al. concluded that "climate change projections based on current general circulation models do not span the range of possibilities consistent with the recent climate record."

There are arguments that the possible range of climate sensitivity and therefore climate responses could be narrower. Specifically, both Yao and Del Genio (1999) and Del Genio and Wolf (2000) had proposed to revise this and to raise the value for the minimum climate sensitivity to a doubling of CO_2 from 1.5°C to 2.0–2.5°C because most GCMs may have incorrectly overemphasized the negative feedbacks from low clouds. Del Genio and Wolf (2000) have found evidence that low clouds get thinner, instead of thicker, with warming (mainly because of the dominant effect of ascent of the cloud base) in the subtropics and midlatitudes. Thinner low clouds with decreasing liquid water path length mean a cloud less capable of reflecting sunlight, which ultimately lessens the impact from the low cloud-temperature cooling feedback carried in most GCMs.

Another scenario that apparently greatly affects climate response is the complex interaction of climate and global carbon cycles. In an extreme case, Cox et al. (2000) proposed a strong positive feedback of global warming that causes a dramatic release of soil organic carbon to the atmosphere. Cox et al. found that the inclusion of such a strong biophysical feedback in a coupled atmosphere-ocean GCM (added with both a dynamic global vegetation and global carbon cycle model) will increase the originally prescribed atmospheric CO_2 from 700 ppm to 980 ppm by the year 2100. This transient numerical experiment predicted a global warming of 5.5 K by 2100, compared with the 4 K scenario without the carbon cycle feedback. The corresponding warming over land is 8 K, instead of 5.5 K without the added atmospheric CO_2 from the strong biophysical feedback. But these authors acknowledged that their results depend critically on the model assumption of a long-term sensitivity of soil respiration to global warming, which may be contradicted by field and laboratory data (Giardina and Ryan 2000).[4]

In contrast, semiempirical estimates by Lindzen (1997) and Idso (1998) that included probable negative feedbacks in the climate system yielded a climate sensitivity of about 0.3 to 0.5 K for a doubling of atmospheric CO_2. Furthermore, Hu et al. (2000) noted that as the sophistication of parameterization of atmospheric convection increases, there is a tendency for climate model sensitivity to variation in atmospheric CO_2 concentration to decrease considerably. In Hu et al.'s study, the change is from a decrease in the averaged tropical surface warming of 3.3 K to 1.6 K for a doubling of CO_2 that is primarily associated with the corresponding decrease in the calculated total atmospheric column increase in water vapor from 29 percent to 14 percent.

The main point that emerges here is that the range of climate sensitivity remains large and it is not sufficiently well quantified either by empirical or theoretical means.

Causes of Recent Climatic Change: Aerosol Forcing

Other recent efforts, such as that of Bengtsson et al. (1999), have highlighted the inconsistency between the differing observed surface and tropospheric temperature trends and simulated GCM trends that try to include forcing factors such as combined anthropogenic GHGs, anthropogenic aerosols (both direct and indirect effects), stratospheric aerosols from the Mount Pinatubo eruption, and changes in the distribution of tropospheric and stratospheric ozone. In addition, Roeckner et al. (1999) have discussed how superposing other forcings, like direct and indirect aerosol effects, on the GHG forcing has led to an unexpected weakening of the intensity of the global hydrologic cycle. We also wish to add that surface or tropospheric warming in combination with lower stratospheric cooling does not uniquely signify a fingerprint of elevated CO_2 concentration. Such a change in temperature lapse rate is also the natural behavior of the atmosphere associated with potential vorticity anomalies in the upper air's flow structure (Hoskins et al. 1985; Liu and Schuurmans 1990). This ambiguity precludes the detection of anthropogenic CO_2 effects without additional, confirming information.

Not all researchers differentiate a forcing by aerosols. For example, Russell et al. (2000) recently cautioned that "one danger of adding aerosols of unknown strength and location is that they can be tuned to give more accurate comparisons with current observations but cover up model deficiencies." Such an important caveat may give a better sense of urgency if we recall that most current GCMs treat the effects of anthropogenic sulphate aerosols by merely rescaling surface albedo according to a precalculated sulphur loading (Räisänen 1999; Roeckner et al. 1999; Covey 2000). Furthermore, at least in the sense of direct radiative forcing, naturally occurring sources like sea-salt and dimethyl sulphide by marine phytoplankton, rather than anthropogenic sources (Haywood et al. 1999; Haywood and Boucher 2000; Jacobson 2001), dominate the variable and inhomogeneous forcing by aerosols. For example, Jacobson (2001) estimated for all-sky conditions that the global direct radiative forcing from combined natural and anthropogenic aerosols is about -1.4 Wm^{-2}, compared with an anthropogenic-only (including the black carbon component) aerosol forcing of -0.1 Wm^{-2}. Haywood and Boucher (2000) stressed that the indirect forcing effect of the modification of cloud albedo by aerosols could range from -0.3 to -1.8 Wm^{-2}, while the additional aerosol influences on cloud liquid water content (hence, precipitation efficiency), cloud thickness, and cloud lifetime are still highly uncertain and difficult to quantify. Therefore, the formulation of an internally consistent approach to determining the climatic effects of CO_2 by including both natural and anthropogenic aerosols in the troposphere remains a critical area of research (Haywood and Boucher 2000; Rodhe et al. 2000).

Nonlinear Dynamical Perspective on Climate Change

A somewhat different interpretation of recent climate change is also possible (Corti et al. 1999; Palmer 1999). In an analysis of Northern Hemispheric 500-mb geopotential heights, the authors showed that the record since the 1950s could essentially be projected in terms of the modes of four naturally occurring, shorter-term, atmospheric circulation regimes, identified in Corti et al. (1999) as Cold-Ocean-Warm-Land (COWL), Pacific North American Oscillation, North Atlantic Oscillation, and Arctic Oscillation patterns. Then, climate variability, viewed as vacillations of these quasi-stationary weather regimes, can be quantified by changes in the probability density function associated with each regime. Palmer and colleagues thus proposed that the impact of anthropogenic CO_2 forcing might be revealed as a projection onto modes of these natural weather regimes. Of course, there is no guarantee that the underlying structure of the weather regimes would remain the same under the perturbation of a different or stronger forcing.

Next, Corti et al. (1999) showed that recent observed changes could be interpreted primarily as an increasing occurrence probability associated with the COWL regime (Wallace et al. 1995), which is caused mainly by the heat capacity contrast between land and sea, perhaps consistent with the projection of the anthropogenic CO_2 forcing. With this idea in mind, the authors proposed to resolve the contentious discrepancy between the rising trend in surface air temperature versus the relative constancy of the lower tropospheric air temperature, as summarized in the NRC (2001) report, the rationale being that most of the recent hemispheric-mean temperature change is associated with the COWL pattern. And since the COWL pattern is primarily a surface phenomenon, one can expect to find a stronger anthropogenic CO_2-forced temperature imprint at the surface than in the troposphere. Above the surface, the land–sea contrast weakens significantly so that no imprint of anthropogenic thermal forcing anomalies persist there. But such a pattern of climatic change—emphasizing surface response over land—seems also consistent with the heat island effect from urbanization, leaving interpretation of the vertical pattern of temperature trend unresolved.

It is, of course, a curious point that no GCM has yet simulated such a vertical pattern of climate change (Bengtsson et al. 1999). The strongest anthropogenic CO_2 response in GCMs is still expected in the middle to high troposphere, simply because of the dominance of direct radiative effects. Santer et al. (2001) confirmed that the wrong fingerprint is observed when compared with that expected from CO_2 forcing of the atmosphere. GCMs simulate trend differences of surface-minus-lower troposphere temperature significantly smaller than the observed results for fifty-one out of fifty-four simulations examined, even for best-effort attempts to account for trend biases introduced by effects of volcanoes and ENSO.

A further question left unanswered by Corti et al. (1999) is why increased CO_2 should lead to an increase in the frequency of the COWL regime. Furthermore, any number of warming influences may contribute to the positive bias of COWL, since the main physical cause of the pattern is the heat capacity contrast between land and sea. In this respect, it is important to point out that the COWL pattern is a robust feature of un-forced numerical climate experiments under various air-sea coupling schemes (Broccoli et al. 1998). But as emphasized by these authors, even though a direct comparison of observations with the model-derived un-forced patterns and changes "has implications for the detection of climate change, [they] do not intend to attribute the recent warming of North-ern Hemisphere land to specific causes."

Broccoli et al. (1998) conclude that separating forced and unforced changes in observational records is difficult. Hence, they focused strictly on pointing out the problem in the methodology introduced by Wallace et al. (1995) by applying the COWL-pattern variability for climate change de-tection. In doing so, they utilized a GCM run forced with CO_2 and tro-pospheric sulphate aerosols to make their points, but they did not elaborate on results with CO_2 forcing alone. Their main conclusion is that the de-composition method of Wallace et al. is not suitable for climate change de-tection because it yields ambiguous results when more than one radiative forcing pattern (like CO_2 and tropospheric sulphate aerosols) is present.

The recognition of climatic change as responses of a nonlinear dy-namical system imposes the strong requirement that GCMs must accurately simulate natural circulation regimes and their associated variabilities down to regional and synoptic scales. This requirement is especially difficult to fulfill because the global radiative forcing of a few Wm^{-2} expected from the anthropogenic CO_2 perturbation is quite small compared with the uncer-tain energy budgets of various components of the climate system, as well as flux errors in model parameterizations of physical processes. For a perspec-tive on the severity of this problem, consider the dynamic phenomenon of midlatitude atmospheric blocking. As part of the AMIP, D'Andrea et al. (1998) have recently confirmed the large differences in blocking behavior produced among the fifteen or sixteen GCMs that span a wide range of modeling techniques and physical parameterizations. When compared with observed blocking statistics, all GCMs showed systematic errors of underestimating both the blocking frequency and the duration of blocking events (almost all models have problems in producing long-lived blocking episodes over the midlatitude Euro-Atlantic and Pacific sectors). Worse still, there is also no clear evidence that high spatial resolution models perform systematically better than low-resolution models. D'Andrea et al. (1998) have thus proposed only ad hoc numerical experiments to study the possi-ble, previously hidden model deficiencies responsible for the large range of GCM performance in simulating atmospheric blocking. Govindan et al.

(2002) highlighted, by performing the detrended fluctuation analysis on daily maximum temperature records for six sites spread across the globe, that seven leading coupled GCMs systematically underestimated the observed long-range persistence of the atmosphere (roughly after time scales longer than two years) and overestimated the daily maximum temperature trend. From that failure of computer models to emulate the observed behavior in the real atmosphere, Govindan et al. deduced that "the anticipated global warming is also [likely] overestimated" by those leading GCMs. Therefore, we conclude that significant challenges in numerical weather and climate modeling remain.

New Observational Scheme

Modeling is but one approach to understanding climate change. To place more confidence in modeling, observational capability must advance. Improved precision, accuracy, and global coverage are all-important requirements. For example, Schneider (1994) has estimated that a globally averaged accuracy of at least 0.5 Wm^{-2} in net solar-IR radiative forcing is required to refine the present unacceptably large range in the estimates of climate sensitivity. In this respect, Goody et al. (1998) have proposed the complementary scheme of interferometric measurements of spectrally resolved thermal radiance and radio occultation measurements of refractivity—with help from Global Positioning System (GPS) satellites—that can achieve a global coverage with an absolute accuracy of 1 cm^{-1} in spectral resolution and 0.1 K in thermal brightness temperature. The resolution capability of 0.1 K is needed to quantify the expected warming from increased greenhouse gases in one decade, while the accuracy of 1 cm^{-1} is needed to resolve differences in possible spectral radiance fingerprints among several causes. Along with a promised high vertical resolution of about 1 km, the complementary thermal radiances and GPS refractivity measurements should produce a better characterization of clouds, since thermal radiance is cloud sensitive but the refraction of GPS radio signals, while sensitive to water vapor and air molecules, is not affected by clouds. These observational schemes thus offer hope to test climate model predictions critically and to detect anthropogenic CO_2 forcing before it would become too large. Additional discussion, including an objective and bias-free strategy to detect global warming due to the enhanced atmospheric concentration of carbon dioxide, about the proposal by Goody et al. (1998) has been elaborated by Keith and Anderson (2001).

CONCLUSION

Our current lack of understanding of the earth's climate system does not allow us to determine reliably the magnitude of the climate change that will be caused by anthropogenic CO_2 emissions, let alone whether this change will be for better or for worse. We raise a point concerning value judgment here because a value assignment is prerequisite to evaluating the need for mitigation of adverse consequences of climate change. If natural and largely *uncontrollable* factors that yield rapid climate change are common, would humans be capable of actively modifying climate for the better? Such a question has been posed and cautiously answered in the negative—as by Kellogg and Schneider (1974). Given current concerns about rapid climate change, several geoengineering proposals are being revived and debated in the literature (e.g., Schneider 1996; Betts 2000; Govindasamy and Caldeira 2000). We argue that even if climate is hypersensitive to small perturbations in radiative forcing, the task of understanding climate processes must still be accomplished *before* any effective action can be taken.

Our review of the literature has shown that GCMs are not sufficiently robust to provide a quantitative understanding of the potential effects of CO_2 on climate necessary for public discussion. Views differ widely on the plausible theoretical expectations of anthropogenic CO_2 effects, ranging from dominant radiative imprints in the upper and middle troposphere (based on GCM results) to nonlinear dynamical responses. Even if a probability could be assigned to a certain catastrophic aspect of CO_2-induced climatic change, this measure can be objective only if all relevant facts, including those factors that are still in the future, are considered in the calculation. Therefore, at the current level of understanding, global environmental change resulting from increasing atmospheric CO_2 is not quantifiable.

Systematic problems in our inability to simulate present-day climate change are worrisome. The perspective from nonlinear dynamics that suggests "confidence in a model used for climate simulation will be increased if the same model is successful when used in a forecasting mode" (IPCC 1990, as quoted in Palmer 1999) also paints a dismal picture of the difficult task ahead. This brief overview shows that we are not ready to tell what the future climate of the earth will look like. The primary reason for our inability to do so is that, even if we have perfect control over how much CO_2 humans introduce into the air, other variable components of the climate system, both internal and external, are not sufficiently well defined. Also, all future climate scenarios performed in various GCMs must be strictly considered only as mere numerical sensitivity experiments, instead of meaningful climate change predictions (Räisänen 1999; Mikolajewicz and Voss 2000). Attempts to integrate the environmental impacts of anthropogenic CO_2 should note limitations in current GCMs and avoid circular logic (Rodhe et al. 2000).

The nonlinear dynamics perspective further allows for the possibility of multiple climate attractors, attractors that may differ in the presence of the thermohaline convection of the ocean. There is also the possibility that infinitesimal forcing might have the potential to cause a sudden transition to a radically different attractor (Posmentier et al. 1999). Practical logistics preclude the use of a GCM for an evaluation of the likelihood of such a sudden transition. But from the point of view of global warming policy, a quantification of the types and probabilities of likely sudden transitions would be essential.

In light of these facts, we support a more inclusive and comprehensive treatment of the CO_2 question, stated as an internally consistent scientific hypothesis, as the rules of science demand. Climate specialists should continue to urge caution in interpreting GCM results and to acknowledge the incomplete state of our current understanding of climate change. Progress will be made only by formulating and testing a falsifiable hypothesis of any predictable effects by the added CO_2.

We present these criticisms with the aim of improving climate model physics and the use of GCMs for climate science research. Furthermore, we are biased in favor of results deduced from observations. We recognize that there are alternative arguments and other interpretations of the current state of GCMs and climatic change (Grassl 2000). In addition, we acknowledge that the arguability of placing the burden of proof for *small* climate impact on those who would maintain or increase the production rate of anthropogenic CO_2. We recommend consulting the IPCC reports (1990, 1996, and 2001) for such a view. On the other hand, as we have just noted, we also realize that the burden of proof for defining an "optimal" world climate is an issue of equal magnitude. Our review points out the enormous scientific difficulties facing the calculation of climatic effects of added CO_2 in a GCM, but we do not claim either to prove or to disprove a significant anthropogenic influence on global climate.

As for a second opinion on our evaluation of climate models' systematic errors and inadequacies (including even the question of added atmospheric carbon dioxide), the thoughts of Bryson (1993) are applicable:

A model is *nothing more* than a formal statement of how the modeler believes that the part of the world of his concern actually works. It may be simple or complex. . . . Because of the size and complexity of the climate system, no set of equations, and thus no mathematical model, has yet been devised to describe or simulate adequately the complete behavior of the system . . . it may be years before computer capacity and *human knowledge* are adequate for a reasonable simulation. This is not to say that present computer simulations are useless—they are simply not terribly good yet. For example, the average error of the largest, most complex GCM simulations of the present rainfall is well over 100 percent. The temperature errors are impressively large also—up to 20 degrees Centigrade for Antarctica, ten degrees in the Arctic, and two to five degrees elsewhere. The main models in use all have similar errors, but it is hardly surprising, since they are all essentially clones of each other.

ACKNOWLEDGMENTS

This is a revised and updated version of an article previously published by Soon et al. (2000), *Climate Research* 18: 259–75. We acknowledge the invaluable contributions of our coauthors, S. Baliunas, S. B. Idso, and K. Ya. Kondratyev. E. S. Posmentier and W. Soon contributed nearly equally to this improved and updated version of this chapter. Neither W. Soon's nor E.S. Posmentier's views represent those of any institutions with which either is affiliated. E. S. Posmentier acknowledges the support of the Long Island University Faculty Research Released Time program.

NOTES

1. Apparently, a fast trigger such as increased atmospheric methane from rapid release of trapped methane hydrates in permafrosts and on continental margins, through changes in temperature of intermediate-depth water (a few hundred meters below sea level), may be one example of a key ingredient for amplification or feedback leading to large climatic change (Kennett et al. 2000; Hinrichs et al. 2003).

2. See a recent economical–statistical critique on the IPCC 2001's SRES by Castles and Henderson (2003) and response by Nakicenovic et al. (2003).

3. Our statement is consistent with the recent retrospective explanation by Wigley and Raper (2002), which concludes that "while emissions scenario changes are the primary reason for the higher projections of global-mean warming in the TAR [IPCC 2001], this should not be interpreted as an indication that aspects of uncertainty in science are of lesser importance. The range of projected warming is as much dependent on uncertainties in the climate sensitivity as it is on emissions uncertainties" (2952).

4. Giardina and Ryan (2000) showed, through data from eighty-two sites over five continents, that over decades the enzyme action of microbes, which ultimately releases carbon dioxide from decomposing soil, is not sensitive to temperature. Instead, it depends more on factors like the availability of nutrients, moisture, soil clay content, and quality of the mineral soil itself. Liski et al. (2003) reemphasized that there is evidence to suggest that decomposition of old soil is rather insensitive to temperature and that litter at later stages of its decomposition "may become more tolerant of climate" (583). In order to explain the relative insensitivity of soil carbon stock to temperature, Thornley and Cannell (2001) hypothesized that "warming may increase the rate of physico-chemical processes which transfer organic carbon to "protected, more stable, soil carbon pools" (592). Hence, it is a distinct possibility that instead of more carbon emission to the atmosphere as temperature rises, more carbon can be stored in the soil carbon system in the long term (where short-term warming is expected to deplete soil carbon).

REFERENCES

Abarca del Rio, R. 1999. The influence of global warming in earth rotation speed. *Ann Geophys* 17:806–11.

Akmaev, R. A., and V. I. Fomichev. 2000. A model estimate of cooling in the mesosphere and lower thermosphere due to the CO_2 increase over the last 3–4 decades. *Geophys Res Lett* 27:2113–16.

Alley, R. 2000. Ice-core evidence of abrupt climate changes. *Proc Natl Acad Sci* 97: 1331–34.

Arthur, W. B. 1999. Complexity and the economy. *Science* 284:107–9.

Bell, J., P. Duffy, C. Covey, L. Sloan, and the CMIP investigators. 2000. Comparison of temperature variability in observations and sixteen climate model simulations. *Geophys Res Lett* 27:261–64.

Bengtsson, L., E. Roeckner, and M. Stendel. 1999. Why is the global warming proceeding much slower than expected? *J Geophys Res* 104:3865–76.

Betts, R. A. 2000. Offset of the potential carbon sink from boreal forestation by decreases in surface albedo. *Nature* 408:1890.

Broccoli, A. J., N.-C. Lau, and M. J. Nath. 1998. The Cold Ocean–Warm Land pattern: Model simulation and relevance to climate change detection. *J Clim* 11:2743–63.

Broecker, W. S. 1987. Unpleasant surprises in the greenhouse? *Nature* 328:123–26.

Bryden, H. L. 1999. Global ocean circulation. In *IUGG XXII General Assembly, Abstract Book,* A1.

Bryson, R. A. 1993. Environment, environmentalists, and global change: A skeptic's evaluation. *New Lit Hist* 24:783–95.

Cai, W., and H. B. Gordon. 1999. Southern high-latitude ocean climate drift in a coupled model. *J Clim* 12:132–46.

Castles, I., and D. Henderson. 2003. The IPCC emission scenarios: An economic–statistical critique. *Energy and Environment* 14:159–85.

Collins, M. 2000a. The El Niño–Southern Oscillation in the second Hadley Centre coupled model and its response to greenhouse warming. *J Clim* 13: 1299–1312.

Collins, M. 2000b. Understanding uncertainties in the response of ENSO to greenhouse warming. *Geophys Res Lett* 27:3509–12.

Corti, S., F. Molteni, and T. N. Palmer. 1999. Signature of recent climate change in frequencies of natural atmospheric circulation regimes. *Nature* 398:799–802.

Covey, C. 2000. Beware the elegance of the number zero: An editorial comment. *Clim Change* 44:409–11.

Covey, C., et al. 2000. The seasonal cycle in coupled ocean-atmosphere general circulation models. *Clim Dyn* 16:775–87.

Cox, P. M., R. A. Betts, C. D. Jones, S. A. Spall, and I. J. Totterdell. 2000. Acceleration of global warming due to carbon-cycle feedbacks in a coupled climate model. *Nature* 408:184–87.

D'Andrea, F., et al. 1998. Northern Hemisphere atmospheric blocking as simulated by 15 atmospheric general circulation models in the period 1979–1988. *Clim Dynamics* 14:385–407.

Davey, M. K., et al. 2002. STOIC: a study of coupled model climatology and variability in tropical ocean regions. *Clim Dynamics* 18:403–20.

Del Genio, A. D., and A. B. Wolf. 2000. The temperature dependence of liquid water path of low clouds in the Southern Great Plains. *J Clim* 13:3465–86.

Delworth, T. L., and K. W. Dixon. 2000. Implications of the recent trend in Arctic/North Atlantic Oscillation for the North Atlantic thermohaline circulation. *J Clim* 13:3721–27.

Dettinger, M. D., and M. Ghil. 1998. Seasonal and interannual variations of atmospheric CO_2 and climate. *Tellus* 50B:1–24.

Dijkstra, H. A., and J. D. Neelin. 1999. Imperfections of the thermohaline circulation: Multiple equilibria and flux correction. *J Clim* 12:1382–92.

Dirmeyer, P. A. 2001. Climate drift in a coupled land-atmosphere model. *J Hydrometeorol* 2:89–102.

Dixon, K. W., T. L. Delworth, M. J. Spelman, and R. J. Stouffer. 1999. The influence of transient surface fluxes on North Atlantic overturning in a coupled GCM climate change experiment. *Geophys Res Lett* 26:2749–52.

Egger, J. 1999. Numerical generation of entropies. *Monthly Weather Review* 127:2211–16.

Essex, C. 1991. What do climate models tell us about global warming? *Pageoph* 135:125–33.

Fedorov, A. V., S. L. Harper, S. G. Philander, B. Winter, and A. Wittenberg. 2003. How Predictable is El Niño? *Bull Am Met Soc* 84:911–19.

Field, C. B., and I. Y. Fung. 1999. The not-so-big U.S. carbon sink. *Science* 285: 544–45.

Fischer H., M. Wahlen, J. Smith, D. Mastroianni, and B. Deck. 1999. Ice core records of atmospheric CO_2 around the last three glacial terminations. *Science* 283:1712–14.

Forest, C. E., M. R. Allen, A. P. Sokolov, and P. H. Stone. 2001. Constraining climate model properties using optimal fingerprint detection methods. *Clim Dynamics* 18:277–95.

Forest C. E., M. R. Allen, P. H. Stone, and A. P. Sokolov. 2000. Constraining uncertainties in climate models using climate change detection techniques. *Geophys Res Lett* 27:569–72.

Gadsden, M. 1998. The north-west Europe data on noctilucent clouds: A survey. *J Atmos Sol-Terr Phys* 60:1163–74.

Gent, P. R. 2001. Will the North Atlantic ocean thermohaline circulation weaken during the 21st century? *Geophys Res Lett* 28:1023–26.

Giardina, C. P., and M. G. Ryan. 2000. Evidence that decomposition rates of organic carbon in mineral soil do not vary with temperature. *Nature* 404:858–61.

Gillett, N. P., M. R. Allen, and S. F. B. Tett. 2000. Modelled and observed variability in atmospheric vertical temperature structure. *Clim Dynamics* 16:49–61.

Glecker, P. J., and B. C. Weare. 1997. Uncertainties in global ocean surface heat flux climatologies derived from ship observations. *J Clim* 10:2764–81.

Goodman, J., and J. Marshall. 1999. A model of decadal middle-latitude atmosphere-ocean coupled modes. *J Clim* 12:621–41.

Goody, R., J. Anderson, and G. North. 1998. Testing climate models: An approach. *Bull Am Meteor Soc* 79:2541–49.

Gordon, C., et al. 2000. The simulation of SST, sea ice extents and ocean heat transports in a version of the Hadley Centre coupled model without flux adjustments. *Clim Dynamics* 16:147–68.

Govindan, R. B., D. Vyushin, A. Bunde, S. Brenner, S. Havlin, and H. J. Schellnhuber. 2002. Global climate models violate scaling of the observed atmospheric variability. *Phys Rev Lett* 89:28501 (4).

Govindasamy, B., and K. Caldeira. 2000. Geoengineering earth's radiation balance to mitigate CO_2-induced climate change. *Geophys Res Lett* 27:2141–44.

Grabowski, W. W. 2000. Cloud microphysics and the tropical climate: Cloud-resolving model perspective. *J Clim* 13:2306–22.

Grassl, H. 2000. Status and improvements of coupled general circulation models. *Science* 288:1991–97.

Grenier, H., H. Le Treut, and T. Fichefet. 2000. Ocean–atmosphere interactions and climate drift in a coupled general circulation model. *Clim Dynamics* 16: 701–17.

Haywood, J. M., and O. Boucher. 2000. Estimates of the direct and indirect radiative forcing due to tropospheric aerosols: A review. *Rev Geophys* 38:513–43.

Haywood, J. M., V. Ramaswamy, and B. J. Soden. 1999. Tropospheric aerosol climate forcing in clear-sky satellite observations over the oceans. *Science* 283: 1299–1303.

Held, I. M., and B. J. Soden. 2000. Water vapor feedback and global warming. *Ann Rev Ene Env* 25:441–75.

Hinrichs, K. U., L. R. Hmelo, and S. P. Sylva. 2003. Molecular fossil record of elevated methane levels in late Pleistocene coastal waters. *Science* 299:1214–17.

Hoffert, M. I., et al. 1999. Energy implications of future stabilization of atmospheric CO_2 content. *Nature* 398:121–26.

Holland, M. M., A. J. Brasket, A. J. Weaver. 2000. The impact of rising atmospheric CO_2 on simulated sea ice induced thermohaline circulation variability. *Geophys Res Lett* 27:1519–22.

Holloway, G., and O. Saenko. 1999. Potholes in the global conveyor: The Arctic interior and the role of everywhere-eddies. *EOS Trans AGU* 80, no. 46: F16.

Hoskins, B. J., M. E. McIntyre, and A. W. Robertson. 1985. On the use and significance of isentropic potential vorticity maps. *Q J Roy Meteorol Soc* 111: 877–946.

Houghton, J. T. 1991. The predictability of weather and climate. *Phil Trans R Soc Lond A* 337:521–72.

Hu, H., R. J. Oglesby, and B. Saltzman. 2000. The relationship between atmospheric water vapor and temperature in the simulations of climate change. *Geophys Res Lett* 27:3513–16.

Huang, H. P., K. M. Weickmann, and C. J. Hsu. 2001. Trend in atmospheric angular momentum in a transient climate change simulation with greenhouse gas and aerosol forcing. *J Clim* 14:1525–34.

Idso, S. B. 1998. CO_2-induced global warming: A skeptic's view of potential climate change. *Clim Res* 10:69–82.

Indermühle, A., et al. 1999. Holocene carbon-cycle dynamics based on CO_2 trapped in ice at Taylor dome, Antarctica. *Nature* 398:121–26.

Intergovernmental Panel on Climate Change (IPCC). 1990. Climate Change: The IPCC Scientific Assessment. Edited by J. T. Houghton et al. Cambridge: Cambridge University Press, 1990.

Intergovernmental Panel on Climate Change (IPCC). 1996. *Climate Change 1995: The Science of Climate Change.* Edited by J. T. Houghton et al. Cambridge: Cambridge University Press.

Intergovernmental Panel on Climate Change (IPCC). 2001. *Climate Change 2001: The Scientific Basis.* Edited by J. T. Houghton et al. Cambridge: Cambridge University Press.

Jacobson, M. Z. 2001. Global direct radiative forcing due to multicomponent anthropogenic and natural aerosols. *J Geophys Res* 106:1551–68.

Jain, S., U. Lall, and M. E. Mann. 1999. Seasonality and interannual variations of northern hemisphere temperature: Equator-to-pole gradient and ocean-land contrast. *J Clim* 12:1086–1100.

Janssens, I. A., et al. 2003. Europe's terrestrial biosphere absorbs 7 to 12 percent of European anthropogenic CO_2 emissions. *Science* 300:1538–42.

Johnson, D. R. 1997. General coldness of climate models and Second Law: Implications for modeling the earth system. *J Clim* 10:2826–46.

Johnson, D. R., A. J. Lenzen, T. H. Zapotocny, and T. K. Schaack. 2000. Numerical uncertainties in the simulation of reversible isentropic processes and entropy conservation. *J Clim* 13:3860–84.

Joos, F., R. Meyer, M. Bruno, and M. Leuenberger. 1999. The variability in the carbon sinks as reconstructed for the last 1000 years. *Geophys Res Lett* 26:1437–40.

Keating, G. M., R. H. Tolson, and M. S. Bradford. 2000. Evidence of long-term global decline in the earth's thermospheric densities apparently related to anthropogenic effects. *Geophys Res Lett* 27:1523–26.

Keith, D. W., and J. G. Anderson. 2001. Accurate spectrally resolved infrared radiance observation from space: Implications for detection of decade-to-century-scale climatic change. *J Clim* 14:979–90.

Kellogg, W. W., and S. H. Schneider. 1974. Climate stabilization: For better or for worse? *Science* 186:1163–72.

Kennett, J. P., K. G. Cannariato, I. L. Hendy, and R. J. Behl. 2000. Carbon isotopic evidence for methane hydrate instability during Quaternary interstadials. *Science* 288:128–33.

Kiehl, J. T., and K. E. Trenberth. 1997. Earth's annual global mean energy budget. *Bull Am Meteor Soc* 78:197–208.

Kirk-Davidoff, D. B., and R. S. Lindzen. 2000. An energy balance model based on potential vorticity homogenization. *J Clim* 13:431–48.

Kodera, K., H. Koide, and H. Yoshimura. 1999. Northern Hemisphere winter circulation associated with the North Atlantic Oscillation and stratospheric polar-night jet. *Geophys Res Lett* 26:443–46.

Kodera, K., and Y. Kuroda. 2000. Tropospheric and stratospheric aspects of the Arctic Oscillation. *Geophys Res Lett* 27:3349–52.

Landsea, C. W., and J. A. Knaff. 2000. How much skill was there in forecasting the very strong 1997–98 El Niño? *Bull Am Meteor Soc* 81:2107–19.

Latif, M., E. Roeckner, U. Mikolajewicz, and R. Voss. 2000. Tropical stabilization of the thermohaline circulation in a greenhouse warming simulation. *J Clim* 13: 1809–13.

Lighthill, J. 1986. The recently recognized failure of predictability in Newtonian dynamics. *Proc R Soc Lond A* 407:35–50.

Lindzen, R. S. 1997. Can increasing carbon dioxide cause climate change? *Proc Natl Acad Sci* 94:8335–42.

Lions, J. L., O. P. Manley, R. Temam, S. Wang. 1997. Physical interpretation of the attractor dimension for the primitive equations of atmospheric circulation. *J Atmos Sci* 54:1137–43.

Liski, J., A. Nissinen, M. Erhard, and O. Taskinen. 2003. Climatic effects on litter decomposition from arctic tundra to tropical rainforest. *Glob Change Bio* 9: 575–84.

Liu, Q., and C. J. E. Schuurmans. 1990. The correlation of tropospheric and stratospheric temperatures and its effect on the detection of climate changes. *Geophys Res Lett* 17:1085–88.

Liu, Z., et al. 1999. Modeling long-term climate changes with equilibrium asynchronous coupling. *Clim Dyn* 15:325–40.

Manabe, S., and R. J. Stouffer. 1993. Century-scale effects of increased CO_2 on the ocean-atmosphere system. *Nature* 364:215–18.

Marotzke, J. 2000. Abrupt climate change and thermohaline circulation: Mechanisms and predictability. *Proc Natl Acad Sci* 97:1347–50.

Marotzke, J., and P. H. Stone. 1995. Atmospheric transports, the thermohaline circulation, and flux adjustments in a simple coupled model. *J Phys Oceanogr* 25:1350–64.

McIntyre, M. E. 1999. How far have we come in understanding the dynamics of the middle atmosphere? In *The 14th European Space Agency (ESA) Symposium on European Rocket and Balloon Programmes and Related Research.* Edited by B. Kladeich-Schürmann. ESA Publications.

Mikolajewicz, U., and R. Voss. 2000. The role of the individual air-sea flux components in CO_2-induced changes of the ocean's circulation and climate. *Clim Dyn* 16:627–42.

Miles, N. L., J. Verlinde, and E. E. Clothiaux. 2000. Cloud droplet size distributions in low-level stratiform clouds. *J Clim* 13:295–311.

Murphy, J. M. 1995. Transient response of the Hadley Centre coupled ocean-atmosphere model to increasing carbon dioxide. Part I: Control climate and flux adjustment. *J Clim* 8:36–56.

Nakicenovic, N., et al. 2003. IPCC SRES revisited: A response. *Energy and Environment* 14:187–214.

National Research Council. 2001. Reconciling observation of global temperature change. Washington, D.C.: National Academy Press.

Oreskes, N., K. Shrader-Frechette, and K. Belitz. 1994. Verification, validation, and confirmation of numerical models in the earth sciences. *Science* 263:641–46.

Palmer, T. N. 1999. A nonlinear dynamical perspective on climate prediction. *J Clim* 12:575–91.

Panagi, M., M. A. Arthur, K. H. Freeman. 1999. Miocene evolution of atmospheric carbon dioxide. *Paleocean* 14:273–92.

Pawson, S., et al. 2000. The GCM-Reality Intercomparison Project for SPARC (GRIPS): Scientific issues and initial results. *Bull Am Meteor Soc* 81:781–96.

Perlwitz, J., H.-F. Graf, and R. Voss. 2000. The leading variability mode of the coupled troposphere-stratosphere winter circulation in different climate regimes. *J Geophys Res* 105:6915–26.

Pielke, R. A., Sr. 2001. Earth system modeling: An integrated assessment tool for environmental studies. In *Present and future of modeling global environmental change: Toward integrated modeling,* 311–37. Edited by T. Matsuno and H. Kida. Tokyo: Terra.

Pitman, A. J., et al. 1999. Key results and implications from phase 1(c) of the Project for Intercomparison of Land-surface Parameterization Schemes. *Clim Dyn* 15:673–84.

Posmentier, E. S., W. Soon, and S. Baliunas. 1999. Natural variability in an ocean-atmosphere climate model. *J Fizik* (Physics) *Malaysia* 19:157–72.

Priem, H. A. 1997. CO_2 and climate: A geologist's view. *Space Sci Rev* 81:173–98.

Rahmstorf, S. 1999. Shifting seas in the greenhouse? *Nature* 399:523–24.

Rahmstorf, S. 2000. The thermohaline ocean circulation: A system with dangerous thresholds? An editorial comment. *Clim Change* 46:247–56.

Rahmstorf, S., and A. Ganopolski. 1999. Long-term global warming scenarios computed with an efficient coupled climate model. *Clim Change* 43:353–67.

Räisänen, J. 1999. Internal variability as a cause of qualitative intermodel disagreement on anthropogenic climate changes. *Theor Appl Climatol* 64:1–13.

Rasch, P. J., and D. L. Williamson. 1990. Computational aspects of moisture transport in global models of the atmosphere. *Q J Roy Meteorol Soc* 116:1071–90.

Rayner, P. J., I. G. Enting, R. J. Francey, and R. Langenfelds. 1999. Reconstructing the recent carbon cycle from atmospheric CO_2 and d ^{13}C, and O_2/N_2 observations. *Tellus* 51B:213–32.

Rodhe, H., R. J. Charlson, and T. L. Anderson. 2000. Avoiding circular logic in climate modeling: An editorial essay. *Clim Change* 44:419–22.

Roeckner, E., L. Bengtsson, J. Feichter, J. Lelieveld, and H. Rodhe. 1999. Transient climate change simulations with a coupled atmosphere-ocean GCM including the tropospheric sulfur cycle. *J Clim* 12:3004–32.

Rotstayn, L. D. 1999. Climate sensitivity of the CSRIO GCM: Effect of cloud modeling assumptions. *J Clim* 12:334–56.

Russell, G. L., et al. 2000. Comparison of model and observed regional temperature changes during the past 40 years. *J Geophys Res* 105:14891–98.

Russell, G. L., and D. Rind. 1999. Response to CO_2 transient increases in the GISS coupled model: Regional coolings in a warming climate. *J Clim* 12:531–37.

Santer, B. D., et al. 2001. Accounting for the effects of volcanoes and ENSO in comparisons of modeled and observed temperature trends. *J Geophys Res* 106:28033–59.

Schimel, D., et al. 2000. Contribution of increasing CO_2 and climate to carbon storage by ecosystems in the United States. *Science* 287:2004–6.

Schmittner, A., and T. F. Stocker. 1999. The stability of the thermohaline circulation in global warming experiments. *J Clim* 12:1117–33.

Schneider, E. K. 2002. Understanding differences between equatorial Pacific as simulated by two coupled GCMs. *J Clim* 15:449–69.

Schneider, E. K., B. P. Kirtman, and R. S. Lindzen. 1999. Tropospheric water vapor and climate sensitivity. *J Atmos Sci* 56:1649–58.

Schneider, S. H. 1994. Detecting climatic change signals: Are there any "Fingerprints"? *Science* 263:341–47.

Schneider, S. H. 1996. Geoengineering: Could—or should—we do it? *Clim Change* 33:291–302.

Sen, A. K. 1986. Prediction and economic theory. *Proc R Soc Lond A* 407:3–23.

Senior, C. A. 1999. Comparison of mechanisms of cloud-climate feedbacks in GCMs. *J Clim* 12:1480–89.

Shutts, G. J., J. S. A. Green. 1978. Mechanisms and models of climatic change. *Nature* 276:339–42.

Soden, B. J. 2000. The sensitivity of the tropical hydrological cycle to ENSO. *J Clim* 13:538–49.

Soon, W., S. Baliunas, C. Idso, S. Idso, and D. R. Legates. 2003. Environmental effects of increased atmospheric carbon dioxide. *Energy and Environment* 14: 233–96.

Soon, W., S. Baliunas, A. B. Robinson, and Z. W. Robinson. 1999. Environmental effects of increased atmospheric carbon dioxide. *Clim Res* 13:149–64.

Stocker, T. F., and O. Marchal. 2000. Abrupt climate change in the computer: Is it real? *Proc Natl Acad Sci* 97:1362–65.

Stouffer, R. J., and S. Manabe. 1999. Response of a coupled ocean-atmosphere model to increasing atmospheric carbon dioxide: Sensitivity to the rate of increase. *J Clim* 12:2224–37.

Sun, D. Z., C. Covey, and R. S. Lindzen. 2000. Vertical correlations of water vapor in GCMs. *Geophys Res Lett* 28:259–62.

Sun, S., and R. Bleck. 2001. Atlantic thermohaline circulation and its response to increasing CO_2 in a coupled atmosphere-ocean model. *Geophys Res Lett* 28:4223–26.

Tanaka, H. L., and H. Tokinaga. 2002. Baroclinic instability in high latitudes induced by polar vortex: A connection to the Arctic Oscillation. *J Atmos Sci* 59:69–82.

Thomas, G. E. 1996. Is the polar mesosphere the miner's canary of global change? *Adv Space Res* 18(3):149–58.

Thornley, J. H. M., and M. G. R. Cannell. 2001. Soil carbon storage response to temperature: An hypothesis. *Annals of Bot* 87:591–98.

Timmermann, A. 1999. Detecting the nonstationary response of ENSO to greenhouse warming. *J Atmos Sci* 56:2313–25.

Tucker, W. 1999. The Lorenz attractor exists. *C R Acad Sci (Série I)* 328:1197–1202.

Valentini, R., et al. 2000. Respiration as the main determinant of carbon balance in European forests. *Nature* 404:861–65.

Victor, D. G. 1998. Strategies for cutting carbon. *Nature* 395:837–38.

Wallace, J. M., Y. Zhang, and J. A. Renwick. 1995. Dynamic contribution to hemispheric mean temperature trends. *Science* 270:780–83.

Watanabe, O., J. Jouzel, S. Johnsen, F. Parrenin, H. Shoji, and N. Yoshida. 2003. Homogeneous climate variability across East Antarctica over the past three glacial cycles. *Nature* 422:509–12.

Wielicki, B. A., et al. 2002. Evidence for large decadal variability in the tropical mean radiative energy budget. *Science* 295:841–44. See additional exchanges in *Science* 296:2095a.

Wigley, T. M. L., and S. C. B. Raper. 2002. Reasons for larger warming projections in the IPCC Third Assessment Report. *J Clim* 15:2945–52.

Wild, M., A. Ohmura, H. Gilgen, J. J. Morcrette, and A. Slingo. 2001. Evaluation of downward longwave radiation in General Circulation Models. *J Clim* 14:3227–39.

Wood, R. A., A. B. Keen, J. F. B. Mitchell, and J. M. Gregory. 1999. Changing spatial structure of the thermohaline circulation in response to atmospheric CO_2 forcing in a climate model. *Nature* 399:572–75.

Yang, F., M. E. Schlesinger, and E. Rozanov. 2000. Description and performance of the UIUC 24-layer stratosphere/troposphere general-circulation model. *J Geophys Res* 105:17925–54.

Yang, X., and M. Wang. 2000. Monsoon ecosystems control on atmospheric CO_2 interannual variability: Inferred from a significant positive correlation between year-to-year changes in land precipitation and atmospheric CO_2 growth rate. *Geophys Res Lett* 27:1671–74.

Yao, M.-S., and A. D. Del Genio. 1999. Effects of cloud parameterization on the simulation of climate changes in the GISS GCM. *J Clim* 12:761–79.

Yu, J.-Y., and C. R. Mechoso. 1999. A discussion on the errors in the surface heat fluxes simulated by a coupled GCM. *J Clim* 12:416–26.

Zorita, E., and F. González-Rouco. 2000. Disagreement between predictions of the future behavior of the Arctic Oscillation as simulated in two different climate models: Implications for global warming. *Geophys Res Lett* 27:1755–58.

INDEX

ABOUT THE CONTRIBUTORS

Robert C. Balling Jr., Arizona State University

Richard C. Balling is a professor in the Department of Geography at Arizona State University and former director of the Office of Climatology at that institution. He has published over a hundred articles in professional scientific journals, lectured throughout the United States and more than a dozen foreign countries, and appeared in several scientific documentaries and news features. He serves as a climate consultant to the United Nations Environment Program, the World Climate Program, the World Meteorological Organization, the United Nations Education, Scientific, and Cultural Organization, and the Intergovernmental Panel on Climate Change. His books include *The Heated Debate: Greenhouse Predictions versus Climate Reality, Desertification and Climate,* and *Satanic Gases: Clearing the Air about Global Warming.* He received his A.B. in 1974 from Wittenberg University, his M.A. in 1975 from Bowling Green State University, and his Ph.D. in 1979 from the University of Oklahoma.

Sallie Baliunas, Harvard-Smithsonian Center for Astrophysics

Dr. Sallie Baliunas is an astrophysicist and has served as deputy director of Mount Wilson Observatory. Her research interests include solar variability and other factors in climate change, magnetohydrodynamics of the sun and sunlike stars, exoplanets and the use of laser electro-optics for the correction of turbulence due to the earth's atmosphere in astronomical images.

Randall S. Cerveny, Arizona State University

Randall S. Cerveny is an associate professor of geography at Arizona State University and creator of the ASU Storm Chasers. His research has been published in *Journal of Geoscience Education, Science, Paleogeography, Paleoclimatology, Paleoecology, Climatic Change, Journal of Geophysical Letters, Nature,* and *Bulletin of the American Meteorological Society.* He is also a contributing editor to *Weatherwise.* He is the recipient of The Antarctica Service Medal of the United States of America and the British Cartographic Society Medal. He received his B.S., M.A., and Ph.D. from the University of Nebraska (1981, 1983, 1986).

John Christy, University of Alabama–Huntsville

Dr. John Christy is a professor of atmospheric science and director of the Earth System Science Center at the University of Alabama in Huntsville where he began studying global climate issues in 1987. In November 2000, Governor Don Siegelman appointed him Alabama's State Climatologist. In 1989 Dr. Roy W. Spencer (then a NASA/Marshall scientist and now a Principle Research Scientist at UAH) and Christy developed a global temperature data set from microwave data observed from satellites beginning in 1979. For this achievement, the Spencer-Christy team was awarded NASA's Medal for Exceptional Scientific Achievement in 1991. In 1996, they were selected to receive a Special Award by the American Meteorological Society "for developing a global, precise record of earth's temperature from operational polar-orbiting satellites, fundamentally advancing our ability to monitor climate." In January 2002 Christy was inducted as a Fellow of the American Meteorological Society. Dr. Christy has served as a contributor (1992, 1994 and 1996) and lead author (2001) for the U.N. reports by the Intergovernmental Panel on Climate Change in which the satellite temperatures were included as a high-quality data set for studying global climate change. He has or is serving on five National Research Council panels or committees and has performed research funded by NASA, NOAA, DOE, DOT, and the State of Alabama and has published many articles including studies appearing in *Science, Nature, Journal of Climate,* and *The Journal of Geophysical Research.* Dr. Christy has provided testimony to several congressional committees. He received M.S. and Ph.D. degrees in atmospheric sciences from the University of Illinois (1984, 1987). Prior to this career path, he had graduated from the California State University in Fresno (B.A,. mathematics, 1973) and taught physics and chemistry as a missionary teacher in Nyeri, Kenya, for two years.

Robert E. Davis, University of Virginia

Robert E. Davis is an associate professor of climatology in the University of Virginia's Department of Environmental Sciences, with primary interests in large-scale (synoptic) climatology, climate change, and bioclimatology. Professor Davis has been a member of the Expert Panel on Global Change Research Strategy for the Environmental Protection Agency, a member of the Data Management Advisory Panel of the National Oceanic and Atmospheric Administration, a contributor to the 1995 Report of the Intergovernmental Panel on Climate Change and has testified on global climate change before the House Science Committee. Editor of *Climate Research: Interactions of Climate with Organisms, Ecosystems, and Human Societies*, he is past president of the Climatology Specialty Group of the Association of American Geographers and past chair of the American Meteorological Society's Committee on Biometeorology and Aerobiology. His scholarly articles have appeared in the *International Journal of Biometeorology, Environmental Health Perspectives, Climate Research, Geophysical Research Letters, Atmospheric Environment, American Journal of Enology and Viticulture,* and *International Journal of Climatology.* He received his Ph.D. in climatology from the University of Delaware.

Oliver W. Frauenfeld, University of Colorado

Oliver W. Frauenfeld is a research scientist with the Cooperative Institute for Research in Environmental Sciences at the University of Colorado. His research focuses on a broad range of topics in climate variability and climate change, with a new emphasis on changes in high-latitude and high-altitude environments. Specifically, this work encompasses the interactions between frozen ground and other cryospheric variables and the overlying atmosphere. He also continues to explore ocean-atmosphere interactions in both the tropics and midlatitudes, such as interactions between the Pacific Ocean and synoptic-scale atmospheric circulation variability of the Northern Hemisphere. His research has been published in the *Journal of Climate,* the *Journal of Geophysical Research, Geophysical Research Letters,* the *International Journal of Climatology,* and *Climate Research.* He was a DuPont Fellow at the University of Virginia in 2001–2002 as well as a Dean's Fellow in 2000–2001. He holds a B.A., M.S., and Ph.D. in Environmental Sciences from the University of Virginia (1995, 1999, 2003).

David R. Legates, University of Delaware

David R. Legates is an associate professor of climatology and director of the Center for Climatic Research at the University of Delaware. His research focuses on hydroclimatology, precipitation, and global change. Working closely with Dan Leathers and Tracy DeLiberty, Legates has established the Delaware Environmental Observing System (DEOS)—a comprehensive, real-time system dedicated to monitoring environmental conditions throughout Delaware. Legates recently testified before the U.S. Senate Committee on the Environment and Public Works about the implications of a paper he coauthored with researchers from the Harvard-Smithsonian Center for Astrophysics that examined the climate variations over the past 1,000 years. He is also the Delaware State Climatologist and associate director of the Delaware Space Grant Consortium, sponsored by NASA. Legates received B.S. degrees in mathematics and geography, an M.S. degree in geography-climatology, and a Ph.D. degree in climatology, all from the University of Delaware. He has taught at the University of Oklahoma and Louisiana State University.

Ross McKitrick, University of Guelph

Professor McKitrick holds a B.A. in economics from Queen's University and an M.A. and Ph.D. in economics from the University of British Columbia. He was appointed assistant professor in the Department of Economics at the University of Guelph in 1996 and associate professor in 2000. His area of specialization is environmental economics and policy analysis. He has published scholarly articles in *The Journal of Environmental Economics and Management, Economic Modeling, The Canadian Journal of Economics, Environmental and Resource Economics,* and other journals, as well as commentaries in newspapers and other public forums. He is co-author of *Taken by Storm: The Troubled Science, Policy, and Politics of Global Warming.* His current research areas include empirical modeling of the relationship between economic growth and pollution emissions; the impact of economic activity on the measurement of surface temperatures; and the climate change policy debate. Professor McKitrick has made invited academic presentations in Canada, the United States, and Europe, as well as professional briefings to the Canadian Parliamentary Finance Committee and to government staff at the U.S. Congress and Senate. He is also a Senior Fellow of the Fraser Institute in Vancouver, B.C.

Patrick J. Michaels, University of Virginia

Patrick J. Michaels is a research professor of environmental sciences at the University of Virginia. He is also a Senior Fellow in Environmental Studies at the CATO Institute and a visiting scientist with the George Marshall Institute. He is past president of the American Association of State

Climatologists and was program chair for the Committee on Applied Climatology of the American Meteorological Society. He holds A.B. and S.M. degrees in biological sciences and plant ecology from the University of Chicago, and he received a Ph.D. in ecological climatology from the University of Wisconsin at Madison in 1979. He is the author of *Meltdown: The Predictable Distortion of Global Warming by Scientists, Politicians, and the Media* and co-author of *Satanic Gases.* Michaels is a contributing author and reviewer of the United Nations Intergovernmental Panel on Climate Change. His writing has been published in the major scientific journals, including *Climate Research, Climatic Change, Geophysical Research Letters, Journal of Climate, Nature,* and *Science,* as well as in popular serials such as the *Washington Post, Washington Times, Los Angeles Times, USA Today, Houston Chronicle,* and *Journal of Commerce.* He has appeared on ABC, NPR's "All Things Considered," PBS, Fox News Channel, CNN, MSNBC, CNBC, BBC, and Voice of America.

Eric S. Posmentier, Dartmouth College

Eric Posmentier is visiting professor of earth science at Dartmouth College, where he teaches atmospheric and marine sciences and related courses. He previously served as chair of marine sciences and chair of physics and mathematics at Long Island University. His wide-ranging research has been supported by university contracts from the NSF, NASA, ONR, and industry groups; he has also been a consultant for NASA and industry. His work has been published in numerous journals, and he served two terms as associate editor of the *Journal of Geophysical Research.* Among his current interests are sudden climate change and the modeling and climatologic interpretation of stable isotopes in precipitation.

Willie Soon, Harvard-Smithsonian Center for Astrophysics

Willie Soon is a physicist at the Solar and Stellar Physics Division of the Harvard-Smithsonian Center for Astrophysics and an astronomer at the Mount Wilson Observatory. He also serves as a receiving editor for *New Astronomy* and a regular contributor of TechCentralStation.com. He writes and lectures both professionally and publicly on important issues related to the sun and other stars, global warming, and the earth, as well as general science topics in astronomy and physics. The multidisciplinary nature of his research range has led to many fruitful collaborations with atmospheric, agricultural, and health scientists in addition to astronomers, mathematicians, space physicists, and oceanographers. He is the co-author of *The Maunder Minimum and the Variable Sun-Earth Connection.*